GERAINT
FRANKLIN

high-tech britain

BUILDINGS OF THE SPACE AGE

WITH PHOTOGRAPHY BY JOHN EAST

BATSFORD

contents

0 | introduction

high-tech britain

WHAT IS HIGH-TECH?

Along with the Sony Walkman, the Commodore 64 home computer and the sounds of the Roland TR-808 drum machine, the shiny surfaces and big spans of high-tech buildings provide a zeitgeisty image of 1980s modernity. High-tech treated architecture almost as a branch of product design, taking cues and technologies from cameras or yachts. It was equally inspired by heroic Victorian engineering and a futurist strain of science fiction that embraces H G Wells, Arthur C Clarke and Frank Hampson's Dan Dare. Some high-tech protagonists spent a childhood tinkering with the construction set Meccano. In its cultural origins as well as its protagonists high-tech was hailed as a recognizably British phenomenon, interpreted variously as a compensation mechanism for Thatcherite deindustrialization or as post-imperial nostalgia for an era when British engineers bestrode the globe.

What is high-tech architecture? In its canonical form, it is characterized by prefabricated and exposed structures, sophisticated servicing and what the art critic Herbert Read described as the 'machine aesthetic'.[1] High-tech is an anatomical architecture of bones and guts. A contrasting line of skin buildings, including Foster's **Willis Faber & Dumas** and **Sapa Factory**, conceal their workings within a smooth, hermetic shell, like a chassis or a computer case. In his 1988 study, Colin Davies offers less a succinct definition than a grab bag of characteristics:

> *we can simply say that its characteristic materials are metal and glass, that it purports to adhere to a strict code of honesty of expression, that it usually embodies ideas about industrial production, that it uses industries other than the building industry as sources both of technology and imagery, and that it puts a high priority on flexibility of use.*[2]

To these could be added the values of lightness (the architect Richard Horden coined the phrase 'Light-Tech'), performance and attitudes to change sometimes characterized as 'loose fit' or 'open-ended'.

High-tech has roots in 19th-century engineering and contemporaneous ideas of moral agency in architecture. If it follows Pugin's dictum that 'all ornament should consist of enrichment of the essential construction of the building', then it shares his ambiguity about whether this is a reductive process (stripping back to the essentials) or an additive one (embellishing construction).[3] Either way, high-tech architects were accused of elevating means to the status of ends, going straight to the question of how without passing why. At best, this gave rise to impeccably detailed, problem-solving buildings; at worst, smart but empty-headed ones. 'Technology is the answer,' said Cedric Price, 'but what was the question?'[4]

In one sense high-tech was a reaction against the wet, sludgy mixes and recalcitrant contractors with whom Richard Rogers and Norman Foster struggled early on. Their generation felt a kinship with earlier advocates of 'light and dry' construction such as the Hertfordshire schools architect Bruce Martin, who urged that 'we must build lightly for

Pages 6–7: Centre Georges Pompidou, Paris, 1972-7, Piano + Rogers.

a life of free and changing activity'.[5] The hard-edged precision of site-ready metals, structural glass, plastics and other synthetic materials was a marvel after an unleavened diet of raw concrete, which by the 1980s was tainted by association with the architecture of the welfare state. Yet high-tech has a surprising amount in common with the preceding New Brutalist movement, including manifest structure, a substantial carbon footprint and a feisty advocate in the form of the historian and critic Reyner Banham.

High-tech could equally be regarded as a counterpart to post-modern (PoMo) architecture. Both emerged in the 1970s in response to a widespread disillusionment with the modern movement. Although their sources varied – PoMo took from high and low culture, high-tech derived imagery from machines or vehicles – they shared an acceptance of modernity and an urge to communicate. High-tech returned to modernism's DNA but came up with something like a genetically modified variant. Instead of functionalism came a 'loose-fit' attitude to patterns of use; the *promenade architecturale* was replaced with a grid of neutral and undifferentiated space and a pure or abstract aesthetic was transformed by treating structure as a source of ornament. Like other 'late' styles such as mannerism or perpendicular, high-tech ramped up the volume through overemphasis, contradiction or repetition. These tactics are typified by the Centre Pompidou's contrast between jumbo tubes in primary colours and an attenuated frame.

High-tech architecture was a child of the Cold War. In the West, material affluence and existential threat were inextricably linked after 1945. It was in response to the arms race and its extension into space that technocratic governments supported programmes of research and development in the defence, electronics, aerospace, communications and synthetic chemicals industries. Spin-offs reached the citizen-consumer in the form of domestic electronics, disposables and improved transportation. For Reyner Banham, these gadgets characterized the 'second machine age'.[6] High-tech architecture was closely bound up with trickle-down technology. Early clients included both high-technology start-ups and corporates. Architects attempted to bypass a sluggish building industry by borrowing technology directly from these sectors. Above all, the imagery of technology was celebrated, be it Olivetti calculators or NASA's Vehicle Assembly Building.

In the 1960s, the noun and adjective 'high-technology', soon abbreviated to 'high-tech', became a shorthand for these technologically advanced sectors.[7] This usage gradually crept into architectural criticism: in 1972 the architect Robert Maxwell described Richard Rogers' work as 'an assembly of high-technology components'.[8] In 1978 the American journal *Progressive Architecture* profiled the **Herman Miller Factory** and the **Hopkins House** under the high-tech banner alongside US projects by Helmut Jahn and others.[9]
What cemented the term was not a polemic but a 'look book' named *High Tech:*

High-Tech *(1978) by Joan Kron and Suzanne Slesin.*

the industrial style and source book for the home (1978). Its authors, the American journalists Joan Kron and Suzanne Slesin, demonstrated how techno-chic interiors could be assembled from industrial-ware, an idea taken up the following year at 'High-Tech', a show at London's Building Centre.[10]

For Kron and Slesin, high-tech was a portmanteau of high style and technology; the term, they explained, was 'currently used in architectural circles to describe buildings incorporating prefabricated or "off-the-rack" building components [or] buildings with a technological look'.[11] Historians devised labels of their own: Kenneth Frampton proposed 'productivism', drawing upon Russian Constructivism, while Charles Jencks coined 'slick tech' and the 'second machine aesthetic', the last in honour of Reyner Banham, his former PhD tutor.[12] Architects, fearing that being pigeonholed was bad for business, tended to dismiss 'high-tech' and the like as journalistic froth.

While it could claim no manifesto and a scant theoretical basis, high-tech architecture nevertheless had its champions and critics, including Reyner Banham, Martin Pawley, John Winter and Peter Buchanan. Colin Davies's *High Tech Architecture* (1988) drew on earlier essays for *Architects' Journal* and *Architectural Review*, illustrated with vibrant colour photography by Martin Charles, Peter Cook, Richard Einzig and the team of Jo Reid and John Peck. Further outlets were *Blueprint*, founded in 1983 by Peter Murray and Deyan Sudjic as an all-encompassing design magazine, and the 'Architecture in Detail' series established by Maritz Vandenberg. Colin Amery was supportive as the *Financial Times*' architecture critic as well as an adviser to the Sainsbury family and others. The FT's annual award for industrial architecture ran for three decades from 1967, starting with Team 4's Reliance Controls building in Swindon (described by the jury as 'so beautifully direct that it looks like a lost vernacular')[13]. 'Foster Rogers Stirling', a 1986 exhibition at the Royal Academy of Arts, brought high-tech before a new audience, as did open design competitions (notoriously that for the National Gallery extension in 1982) and international expos. *Building Sights*, four series of 'architectural shorts' that aired on BBC Two between 1988 and 1996, featured Zaha Hadid at the **Willis Faber & Dumas** building, Will Alsop taking a bath at the **Hauer-King House** and Norman Foster walking on the wings of a Boeing 747.

Reyner Banham died before he was able to complete his projected book, *Making Architecture: The Paradoxes of High-Tech*. He intended to unpick the contradictions of an architecture poised between art and science, monumentality and transience, and the symbolic and literal aspects of technology.[14] At its core is a dilemma between the substance of advanced technology and its looks; between performance and the performance of performance. This sense of rhetoric, knowingness or artifice sets high-tech apart from the notionally 'pure' engineering structures that form its reference points. On this basis, technology is to high-tech what science is to sci-fi.

ANTECEDENTS AND PIONEERS

What were the antecedents of high-tech architecture? Architects such as Nicholas Grimshaw sought to locate themselves within a frame of reference that took in 19th-century engineering structures such as the Palm House at Kew Gardens (1848, Decimus Burton with Richard Turner), Crystal Palace (1851, Joseph Paxton with Fox, Henderson & Co) and the Galerie des Machines in Paris (1889, Ferdinand Dutert with Victor Contamin). These iron and glass structures had already been cast as pathfinders of the modern movement by the historians Siegfried Giedion and Nikolaus Pevsner. In the 20th century, innovation in building technology occurred on the fringes of architectural culture, through a series of improvisations and appropriations wrought by what Will McLean has termed the 'aberrant practitioner'.[15] Aspects of high-tech architecture are prefigured by a clutch of radical inter-war projects, mostly unrealized, including Vladimir Shukhov's Shabolovka

The Crystal House, designed by George Fred Keck for the 1934 'Century of Progress' International Exposition in Chicago.

Tower in Moscow (1920–2); the Constructivist schemes of Leonid, Viktor and Alexander Vesnin; Heinz and Bodo Rasch's Suspension Houses (1927–30); George Fred Keck's Crystal House (1934); Oscar Nitzchke's Maison de la Publicité (1934–6) and Jaromír Krejcar's Czech Pavilion at the Paris Expo of 1937.

If there was a patron saint of technological disruption, it would be R Buckminster Fuller, a self-taught designer and inventor who assumed the status of a technological prophet in the 1960s. Derived from the techniques of aircraft production, his Dymaxion House (1927, redesigned in 1944–6) was a prototype aluminium 'dwelling machine' that was low-cost, mass-produced and expendable, condensing architecture into product life cycles. High-tech architects loved Bucky's principle that architecture could be reckoned by measurable performance criteria – hence his famous enquiry, 'How much does your building weigh?'[16] The idea of 'doing more with less' was illustrated by Fuller's adoption, from 1948, of the geodesic dome, a lightweight lattice shell of interlocking triangles that was capable, he argued, of being scaled up to a 3km (1.9-mile) biosphere over Manhattan.[17] *Operating Manual for Spaceship Earth* (1969) brought together Fuller's universal yet idiosyncratic theories on energy, sustainability and systems theory.[18] He collaborated with Norman Foster on projects for the Samuel Beckett Theatre for St Peter's College, Oxford (1971), Climatroffice (1971) and Autonomous House (1982).

Other pioneers quietly advanced prefabrication and industrialized building. Jean Prouvé was a metalworker and *constructeur* who progressed to furniture, demountable steel structures and building kits based on components fabricated in his workshops outside Nancy, north-eastern France. At the Aéro-Club Roland-Garros at Buc (1935–6) and the Maison du Peuple at Clichy, Paris (1937–9), both realized with architects Eugène

The Montreal Biosphere, Quebec, 1965–6 by R Buckminster Fuller. Built as the US pavilion at Expo 67, it is now an environmental museum.

Beaudouin and Marcel Lods, thin panels of sheet metal were pressed and folded for rigidity and sealed into joints. Like Fuller, Prouvé prefabricated complete toilet cubicles in the factory, while the latter project, a community centre, incorporated sliding partitions and mechanisms that enabled sections of the first floor and the skylights to be opened out. It was perhaps the promise of total flexibility that swayed Prouvé's decision as chairman of the competition jury for the Centre Pompidou.

Like Prouvé, Konrad Wachsmann started out 'on the tools', training as a joiner and carpenter before studying architecture. As chief designer at Christoph & Unmack, a manufacturer of flatpack timber structures, he thought hard about the idea of a kit of modular components. The idea of a universal joint was central to his subsequent work in the United States, including the Packaged House system, developed with Walter Gropius in 1941–2, and the Mobilar system of 1944–5.[19] This was one of several patented systems that refined the space frame, the three-dimensional lattice of interlocking members that had been devised by Alexander Graham Bell almost half a century earlier. Wachsmann's modular thinking was widely influential, informing prefabricated school building systems in England and the US.

Elsewhere, one-off designs were mined for their approach to materials, techniques or aesthetics. This was the explicit intention of the Case Study House programme launched by John Entenza of *Arts & Architecture* magazine in 1945. The contributions of Raphael Soriano, Craig Ellwood and Pierre Koenig were pored over by scores of visiting British architects – including Richard Rogers and Norman Foster – for their combination of mass-produced components and airy informality. Charles and Ray Eames' 1949 house in Los Angeles incorporated catalogue windows and other 'as found' components and made it look easy. With 'just a few chairs and a house', as Peter Smithson explained, the couple found magic in the everyday.[20]

What were the British antecedents? In the late 1950s, a handful of wayward student schemes show a preoccupation with technological imagery. At the Regent Street Polytechnic, Michael Webb achieved infamy with a 'bowellist' project of 1958 for the headquarters of the Furniture Manufacturers' Association, in which an assemblage of tubes and organ-like forms was hung outside a skinny frame. For these Webb specified 'gunite' concrete sprayed over shaped steel mesh reinforcement. Before he discovered classical architecture, John Outram hymned the transformative possibilities of new materials such as plastics and mastics, asserting 'the new technology is the glue technology'.[21] His 1959 project for a rooftop boiler house incorporated a fibreglass monocoque shell, curved-corner windows and neoprene gaskets, anticipating the future schemes of his Architectural Association schoolmate Richard Rogers.

Several student schemes were gathered in the first issue of *Archigram*, the most influential of the 'little magazines' purveying radical architecture. It was edited by Peter Cook and David Greene, who soon brought in Ron Herron, Warren Chalk, Dennis Crompton and Michael Webb. Over nine and a half issues, they offered fun, emancipation, mobility and gadgetry in an anarchic, ironic, pop-inspired subversion of the modern movement. David

The Eames House, Pacific Palisades, Los Angeles, 1949 by Charles and Ray Eames.

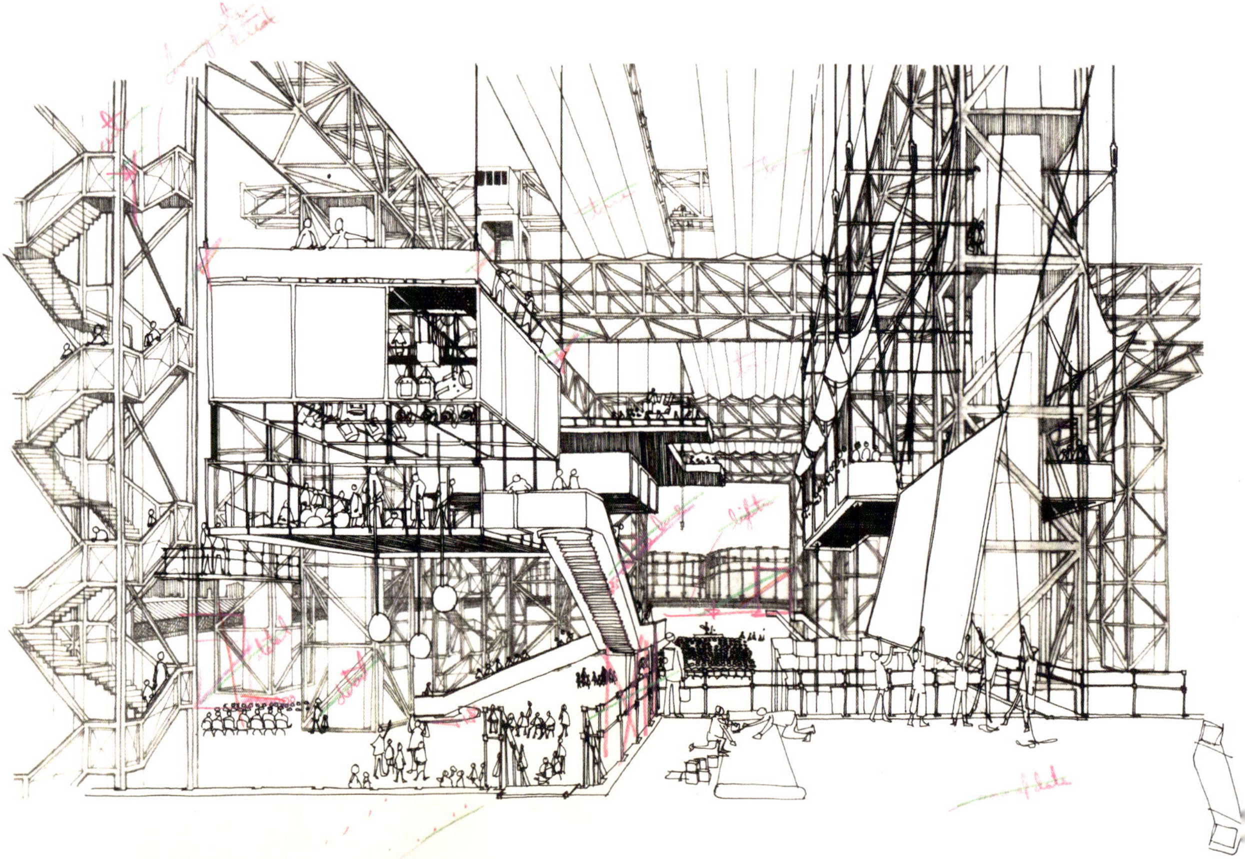

Fun Palace project, c. 1963–4 by Cedric Price.

Greene's poem in *Archigram 1* (1961) conveys a sense of unbounded technological optimism:

you can roll out steel
any length
you can blow up a balloon
any size
you can mould plastic
any shape
blokes that built the forth bridge
THEY DIDN'T WORRY

Over Archigram's nine and a half issues, it is possible trace a dematerializing shift from the concrete to the virtual, from a collectivist to a consumerist sensibility, and from urban networks such as Peter Cook's Plug-in City (1962–4) to capsules and kits that formed interfaces between the human body and its environment.

Just as influential was Cedric Price, who channelled the indeterminacy of John Cage's music into an anti-architecture. His 'Fun Palace' project (c. 1963–4) for Joan Littlewood was no building in the conventional sense but an open-ended, neutral framework into which replaceable modules could be plugged.[22] It was a culture machine for everyone to enjoy. Littlewood recalled being shown Price's initial sketches:

The drawing was almost inexplicable.
I could make out filigree towers, varied areas at different levels, there were galleries, gantries and escalators – it looked airborne.
'Can it be kept clean?'
'It's a self-washing giant.'
'And those things?'
'Moving walkways and catwalks. No, you're pointing at the radial escalators. They can be steered.'[23]

With Price's Potteries Thinkbelt (1964–6) the concept of a network in flux was expanded to encompass the post-industrial landscape of the North Staffordshire Potteries. Price and Archigram were not alone: by the late 1950s urban megastructures were being conceived across the globe, including Yona Friedman's multilayered Spatial City (1958–60); Marine City (1958) by Kiyonori Kikutake, a founder of the Japanese Metabolist movement; New Babylon, a post-revolutionary society conceived between 1959 and 1974 by the situationist Constant Nieuwenhuys; and Takis Zenetos's Electronic Urbanism studies. Utopian pipe dreams these might have been, but they provided the technological optimism and futurist imagery that fuelled high-tech. The argument ran: if these people could dream up whole cities, how difficult would it be to apply that thinking to a single building?

EARLY CONNECTIONS

Architecture in late 20th-century Britain – James Stirling aside – was dominated by the 'big four' high-tech practices of Norman Foster, Richard Rogers, Nicholas Grimshaw and Michael and Patty Hopkins. Having trained at the Architectural Association School of Architecture at Bedford Square (with the notable exception of Norman Foster), their professional and social networks converged on that part of London north of Oxford Street, taking in Marylebone, Fitzrovia, Regent's Park, Camden, Belsize Park and Hampstead. Their practices became finishing schools for a clan of smart young architects who, Peter Cook noted, 'seem to know each other a little too closely and seem to need no outside sustenance'.[24] In 1989 Charlotte Ellis profiled the rising generation that had passed through the offices of Foster, Rogers and Hopkins, noting Richard Horden, Ian Ritchie, Jan Kaplický, Alan Stanton, Birkin Haward and Joanna van Heyningen, David Marks and Julia Barfield, Chris Wilkinson and James Eyre, Alex Lifschutz and Ian Davidson, and Jamie Troughton and John McAslan.[25] Notwithstanding its 'toys for the boys' image, women made decisive if overlooked contributions to British high-tech, starting with Wendy Foster, Su Rogers, Patty Hopkins and Eva Jiřičná, and including engineers such as Jane Wernick, Sophie Le Bourva and Margaret Law.

If this is a family tree, it can be traced back to the 1961 meeting of Richard Rogers and Norman Foster as scholarship students at Yale University. Different in their backgrounds and temperaments, they developed a close working relationship based on rivalry and shared instincts. On their return to London the collaboration continued as Team 4, joined by Wendy Cheeseman and, initially, Wendy's sister Georgie Wolton, the only fully qualified member of the group. Su Rogers was also a core member, as was, in effect, the structural engineer Tony Hunt. Of Team 4's completed buildings the turning point was Reliance Controls (1965, demolished 1991), an electronics factory in Swindon. Its inspirations were Californian: the exposed, welded steel frame and profiled roof deck derived from the Case Study Houses of Raphael Soriano and A Quincy Jones, while the open plan and integrated services and structure pointed to Ezra

Ehrenkrantz's School Construction Systems Development (SCSD) programme, aspects further refined in Foster's 1967 Newport School competition entry.

After Team 4 disbanded, Richard and Su Rogers refined the framed aesthetic at the **Spender House** (1968–9) and the **Rogers House** (1969–70). A radical break came with their competition entry for a Zip-Up House for the American chemical corporation DuPont (1968–9). It had a wrap-around skin of insulated panels, 'zipped' into neoprene gaskets. The panels formed a rigid, self-supporting 'monocoque' shell on the principle of vehicles and aeroplanes. It looked like an extruded tube with glazed ends, round edges and radiused bus windows, perched on pink props like a flamingo. Although unrealized, aspects of the Zip-Up design surfaced at the roof extension of the Design Research Unit's offices in Aybrook Street, London (1969–70, demolished), and at UOP Fragrances, Tadworth, Surrey (1973, demolished), which sported full-height panels of glass-reinforced concrete (GRC).

In 1970 Richard Rogers embarked on a partnership with Renzo Piano, a Genoese architect working on lightweight structures and industrial materials. Their first joint design was the most significant: the *grand projet* of the Centre Georges Pompidou (1972–7) and the adjoining underground IRCAM Institute (1975–8) in the Beaubourg district of Paris. It was the structural engineer Ted Happold who persuaded Rogers to enter the competition. The design was jointly worked up by a small Piano and Rogers team that included Su Rogers, John Young and Marco Goldschmied in conjunction with Happold and members of his 'Structures 3' group at Ove Arup & Partners, including Peter Rice and Tom Barker.

Above: Reliance Controls, Swindon, 1965 to designs by Team 4.
Opposite: Centre Georges Pompidou, Paris, 1972–7 by Piano + Rogers.

Théâtre du Renard
12 Rue du Renard
38 Porte d'Orléans

Pompidou is high-tech at its most radical and didactic. To offer maximum internal flexibility, the structure, services and main circulation were shifted outside, supported by a steel exoskeleton. The play of forces is palpable: hollow tubes are used in compression and vertical rods in tension, connected by cast steel nodes dubbed 'gerberettes'.[26] Overseen by Rice, they resemble the beaks of primordial birds. Happold compared the façade to the Crystal Palace or an oil refinery, but closer in spirit were the 1960s megastructures of Cedric Price, Archigram and the Japanese Metabolists.[27] The initial design offered movable walls and even floors. As an urban proposition that included a new piazza, the Place Georges Pompidou, it was a provocative statement in tune with the countercultural mood that prevailed after the protests of May 1968.

Norman and Wendy Foster founded Foster Associates in 1967. Their early projects established the practice's future preoccupations with uncanny prescience. Their first job, an Operations and Amenity Centre for Fred Olsen Lines in Millwall Docks (1969–70, demolished), brought together office workers and dockers in the same open-plan building with the aim of improving industrial relations. Its end walls were fully glazed in the aluminium and neoprene 'Pittco T wall system' Foster imported from the United States. Offices for Computer Technology Ltd (1971, demolished) were clad in aluminium sandwich panels with curved corners and neoprene seams. Perhaps Foster's most radical building was the most short-lived: an inflatable structure completed within weeks to provide a makeshift office for the firm. Prestige commissions from IBM, **Willis Faber & Dumas** and the **Sainsbury Centre for Visual Arts** led to the Hongkong & Shanghai Bank Headquarters, won at competition in 1979.

The Hongkong & Shanghai Bank Headquarters, Hong Kong, 1982–5 by Foster Associates.

Constructed in 1982–5 as formal negotiations for the handover of Hong Kong to the People's Republic of China got underway, the Hongkong Bank was by far Foster's most complex challenge to date. A management contractor was appointed to coordinate the dozens of international subcontractors who supplied cladding, services and other packages. After considering structural options with Ove Arup & Partners, a suspension structure was chosen. Pairs of steel masts are spanned by five bridges that function as 'sky lobbies', served by high-speed lifts. Triangular trusses spanning between these columns carry hangers that support the floors below. To meet the city's planning ordinances, the tower was divided into three vertical slices with setbacks front and back and services and circulation at the ends. At the base is a ten-storey atrium, lit by sun-scoops with motorized mirrors to track the sun.

Nicholas Grimshaw and Terry Farrell joined forces in 1965, having met during a stint at the London County Council architect's department. Farrell trained at the Newcastle University School of Architecture, where his fifth-year design thesis, a 'Climatron' dome-cum-seaside pier alongside Blackpool Tower, demonstrated interests in Buckminster Fuller alongside the taste cultures of a working-class resort. Grimshaw completed his architectural education at the Architectural Association. The school's 1960s preoccupations with technology, cybernetics and urban networks are seen in his final-year thesis for an 'urban university' in Covent Garden, supervised by Peter Cook and Cedric Price. Farrell/Grimshaw's maiden commission, received in 1965 from Grimshaw's uncle, was for an international students' hostel in Bayswater, west London. Farrell refurbished six terraced houses, while Grimshaw contributed a cylindrical service tower at the rear (since demolished). Prefabricated bathroom pods like those patented by Buckminster Fuller plugged into a helical steel framework, accessible from a continuous ramp.

It was a division of labour that anticipated the pattern of practice over the following 15 years: Grimshaw nurtured a sequence of low-cost industrial buildings, starting with a factory for Rotork Controls in Bath (1971–2) and a 1972–3 warehouse at Runnymede, Surrey, while Farrell oversaw mixed-use urban interventions and social housing. After the partners went their separate ways in 1980, Grimshaw applied his 'kit of parts' approach to larger and more complex programmes. But it was Farrell who Deyan Sudjic heralded as 'the man who took high tech out to play', designing the UK's first Teflon-coated fabric roof for Clifton Nurseries in Covent Garden (1980–1) and adding PoMo roof vents and colour to Peter Rice and Ian Ritchie's Shelterspan construction system for a temporary translucent pavilion at Alexandra Palace, north London (1980–1).[28] His Post-Modern 'air rights' buildings Embankment Place (1987–90) and Alban Gate (1988–92) showcased sophisticated transfer structures engineered by Ove Arup & Partners.

Michael and Patty Hopkins set up in practice in 1976. The two met at the AA, where Patty's final-year project of 1967 was well received by Norman Foster as external critic. Subsequently Michael became a partner in Foster Associates where he oversaw the **IBM Pilot Head Office** before working on **Willis Faber & Dumas**. The **Hopkins House** in Hampstead became a calling card for the new practice in a similar vein to the **Rogers House**, the **Spence and Webster Houses** and Richard Horden's **Yacht House**. Its rigor and discipline was carried across to the industrial projects for **Greene King**, **Patera** Products Ltd and **Schlumberger**. The latter's mast and cable suspension system, supporting a string of tents, represents the culmination of Hopkin's forays into the machine aesthetic before taking a more contextual turn at **Lord's Cricket Ground**.

Who were high-tech's clients? Many commissions came from smart and youthful firms – what would now be termed 'start-ups' – in commerce, manufacturing, computing and finance. The high-technology industries of electronics and microcomputing demanded high levels of precision, hygiene and environmental conditioning, while 'loose-fit' sheds proved ideal for the complex programmes and changing process flows of industrial facilities. 'Progressive' employers aimed to raise productivity and avoid labour disputes by improving facilities and working conditions and, in an industrial context, by dissolving divisions between blue- and white-collar workers. Most attractive to architects were the industrial patrons of good design for whom architecture was an extension of their corporate branding, including such transnational concerns as IBM, Olivetti, Herman Miller and Cummins. Public buildings followed, with the Centre Pompidou and the **Sainsbury Centre** providing precedents for the application of the industrial metaphor to cultural and sports programmes. The commissioning of **Stansted Airport**, the stations of the Jubilee Line Extension, overseen by the architect Roland Paoletti, and the Eurostar terminals at **Waterloo** and later St Pancras confirmed the 'official' acceptance of high-tech not only as an aesthetic but as a model of architect-engineer collaboration.

ENGINEER-ARCHITECT DIALOGUES

As buildings swelled to absorb more functions, services and the demands of the motor car, all kinds of specialists were needed to coordinate these new complexities. Big projects tended to be broken into chunks to be assigned to consultants before being reintegrated into an articulate and rational whole. High-tech architecture could be seen as a visual manifestation of this collaborative process between architects, engineers, manufacturers and contractors. As Andrew Saint shows in *Architect and Engineer: A Study in Sibling Rivalry* (2007), it formed part of a long and dynamic relationship between the professional disciplines.

It was Ove Arup who furthered a new and intimate role for the consulting engineer based on creative support and shared aims. He built Ove Arup & Partners into a global concern capable of coordinating the Sydney Opera House (1959–73), Centre Pompidou and the Hongkong & Shanghai Bank Headquarters. The firm followed the path of increasing professional specialization, branching into civil engineering, building services, acoustics, façade engineering, fire safety engineering, geotechnics and infrastructure planning. Arup Associates was formed in 1963 as a quasi-independent practice in which groups of architects, engineers and others tackled structure, services and composition in the round, on the interdisciplinary model of the American practice Skidmore, Owings & Merrill (SOM).

Arup's sense of engagement with the principles and aesthetics of modern architecture was shared by Felix Samuely. Trained at the Technische Hochschule in Berlin-Charlottenburg, he arrived in England in 1933, making inroads into welded tubular steel, precast and pre-stressed concrete, space structures and folded slab construction. With Powell and Moya he devised the Skylon for the 1951 Festival of Britain, a masted structure that hung in mid-air with no visible means of support (just like the UK economy, it was quipped). After his death his senior partner Frank Newby took over 'Sam's', sustaining a creative dialogue with architects. 'What I really did,' he recalled, 'was give architects confidence that they could design structures.'[29]

But confidence was one thing, hubris another. In 'Hi-tech or Mys-tech', an address to the Royal Institute of British Architects, Newby poured cold water on claims that high-tech architecture was technically innovative, elsewhere observing that:

> *Architects just started using structure as decoration and, because it carried load, we had to deal with it. It made a lot of extra work. But if you are asking me whether high-tech architecture advanced technology in any way, I would have to say no. It made no contribution at all. There was nothing in all of it half as testing as the problems of a long span bridge or a decent sized stadium.*[30]

In his posthumous book *An Engineer Imagines*, Peter Rice cautioned engineers against playing the defensive role of Iago, who in Shakespeare's play uses a reductive rationality to undermine Othello's noble passions.[31] If engineers were to meet architects on their own terms – and he thought they should – then inventiveness, creativity and humanity were needed on both sides. Most of Rice's career was spent at Arup's, although he moonlighted with Renzo Piano before setting up RFR with the designer Martin Francis and the architect Ian Ritchie. Rice enjoyed picking up and turning over discarded technologies, using 'leaves' of ferro-cement and ductile iron in the roof structure of Renzo Piano's Menil Collection (1983–6) in Houston, Texas, and post-tensioned stone for the Pavilion of the Future at Seville Expo '92.

Ted Happold is remembered as 'a great starter of projects, a promoter of ideas, and a builder of teams', first as head of Arup's Structures 3 before founding Buro Happold in 1976.[32] Visiting Riyadh in 1967, Happold met the engineer Frei Otto and the architect Rolf Gutbrod, pioneers of lightweight tensile structures. Otto's ideas, including the use of models as design and testing tools, became a major influence on Happold and the two joined forces regularly and freely. Their collaborations included the Mannheim Bundesgartenschau pavilions (1973–5, with Carlfried Mutschler & Partners), a timber lattice gridshell supporting a free-form fabric membrane; the King Abdulaziz Sports Centre in Jeddah (1976–9), a cable-net structure with twin-skin membranes; and the Munich Aviary (1978–82, with Jörg Gribl) a nebulous tensile structure clad in woven steel mesh. Research into pneumatic structures led to roofs using foil cushion technology at Hopkins' 1991–2 additions to **Schlumberger**.

Tony Hunt shared similar inclinations, commitments and networks with architects of his own generation and was happy to play the same game. Starting out at Samuely's, Hunt established Anthony Hunt Associates (AHA) in 1962. Like Team 4, it was a small, nimble practice with an informal and creative ethos that fostered teamwork. Ted Happold remarked that Hunt was a 'collector' of talents and, like his high-tech collaborators, he nurtured several engineers who in turn established their own reputations and practices, including Mark Whitby (Whitby & Bird, Whitby Wood), Neil Thomas (Atelier One), Matthew Wells (Techniker) and Stephen Morley (SMP, Bianchi-Morley). 'Working with Tony was liberating, and with him, we all went from being nursemaids to magicians,' Whitby recalled.[33] AHA's output was wide-ranging but it was Reliance Controls that pointed the way to Hunt's main contribution: the refinement of a vocabulary for exposed structural steel that exploited developments in welding and casting technologies.

HIGH-TECH INTERNATIONAL

Often typecast as a British movement, high-tech became a global currency, underpinned by Buckminster Fuller, Jean Prouvé and the Japanese Metabolist movement. Although its first urban statements, the Centre Pompidou in Paris and the Hongkong & Shanghai Bank Headquarters, were designed by London-based practices, they were detailed and made by international teams of consultants and subcontractors. National prowess in technology tended to be showcased in international exhibitions such as Expo '70 at Osaka and Expo '92 in Seville.

France enjoyed strong links between architecture and industrial design. In the production-led tradition of Jean Prouvé were Bernard Schoeller's mass-produced Tournesol swimming pool (1969) and François Deslaugiers' Centre Régional d'Informatique in Nemours (1975–9), which was prefabricated from light industrial elements. Culture, technology and monumentality came together in the Parisian *grands projets* promoted by successive presidents. On the heels of the Centre Pompidou came the Cité des Sciences et de l'Industrie at Parc de la Villette (1985–6, Adrien Fainsilber), which transformed an unfinished livestock auction hall with escalators and long-span trusses. Most impressive were the three serres (or greenhouses) of structural frameless glazing, with a pioneering cable support system engineered by Rice Francis Ritchie (RFR). Completed in 1987, the Institut du Monde Arabe, Paris (Jean Nouvel, Gilbert Lezenes and Pierre Soria with Architecture Studio), reconciles the notion of the 'intelligent façade' with the language of cultural pluralism. Its exterior is made up of hundreds of photosensitive apertures that regulate the amount of light entered in the building, reinventing the Islamic *mashrabiya*.

Something of the brittle purity of the early work of Michael and Patty Hopkins is evident at De Fantasie (1982–4), in the Dutch new town of Almere, by the Amsterdam-based practice of Jan Benthem and Mels Crouwel. The winner of a design competition in which the prize was the use of a plot for five years, it was accordingly designed as a glass box sitting lightly on a demountable space-frame base. The same clarity of execution is evident at their premises for Mors Systeemplafonds at Opmeer (1987–8), a tube building made up from pin-jointed, triangulated trusses like those of the **Sainsbury Centre**. An office and showroom share a glass curtain wall that is hung inside the trusses, while the profiled steel cladding of the warehouse is wrapped around the outside.

In West Germany, two mega-projects spanned the 1970s and the territory between functionalism and high-tech. Aachen University Hospital (1969–84, Weber, Brand & Partner) is a 'mat building' in which wards, laboratories, medical facilities and courtyards are criss-crossed by circulation spines and serviced from perimeter towers. Its red steel gantries and elephantine tubes in yellow and silver stripes are as deadpan as Ludwig Leo's Umlauftank in Berlin. At the Internationales Congress Centrum Berlin (1975–9, Ursulina Schüler-Witte and Ralf Schüler), specialized facilities are packed into an architectural freighter sporting external trusses, stair towers and ventilation ducts. Reverting to 19th-century typologies of glasshouses and train sheds is the Glashalle at the Leipziger

De Fantasie, Almere, Netherlands, 1982–4 by Benthem Crouwel Architects.

Messe or Leipzig trade fair (1994–5, Ian Ritchie Architects with von Gerkan, Marg & Partner), a barrel vault of frameless glass hung from arched trusses.

A Swiss affinity for a technological modernism may owe something to traditions of Calvinist puritanism, precision tool making and a civil engineering heritage of bridges, tunnels and hydroelectric dams. Theo Hotz established his reputation with two commissions for the Swiss telecommunications agency PTT. The Fernmeldezentrum Herdern (telephone exchange) at Zürich-Altstetten (1974–8) has a flexible floor plan, stair towers and services becoming expressive elements on the perimeter. At the hulking Postbetriebszentrum (postal sorting office) in Zürich-Mülligen (1981–5), services were integrated into an all-over cladding of aluminium sandwich panels that Hotz likened to a vehicle chassis. The practice later refined a vocabulary of glass in projects such as the Messe Basel (Basel Trade Fair, of 1996–8).

The collaboration between Fritz Haller and Paul Schärer of the furniture manufacturer USM explored the affinities between the assembly of a chair, say, and a building. After developing the Maxi, Midi and Mini kits for prefabricated buildings of different spans and performance requirements, they moved onto the USM Haller system of furniture, based on steel sheets and tubes connected by a ball joint, patented in 1965 and still in production today. Better known as a furniture designer and author, Werner Blaser designed several precocious buildings such as the warehouse and offices for J Lüber GmbH at Märkt, Baden-Württemberg, Germany. A light-framed enclosure is held up with the permanent scaffolding of an external 'Varitec' space frame.

Aachen University Hospital, Germany, 1969–84 by Weber, Brand & Partner.

In the reconstruction of Italy, reinforced concrete was the order of the day; building engineers such as Pier Luigi Nervi, Riccardo Morandi and Aldo Favini derived an expressive architecture from the material. At the church of Mater Misericordiae in Baranzate, Milan (1956–7, Angelo Mangiarotti and Bruno Morassutti with Aldo Favini), a tablecloth of translucent glass is draped over a slab of pre-stressed elements resting on four inset piloti. Approaches to steel were equally catholic. The new flower market in Pescia, won at competition in 1970 by a team led by Leonardo Savioli and completed in 1981, is a mechanistic essay in tension. A fully glazed, column-free trading hall of 10,400m^2 (112,000ft^2) is divided into five bays, hung from pairs of clustered masts.

Italian high-tech is synonymous with Renzo Piano. Born into a family of builders, his early work experimented with assemblies of plastics, steel and ferro-cement. His Italian Industry Pavilion for Expo '70 at Osaka, Japan (1969–70), was a precocious framework of masts and outriggers, from which was hung a puckered membrane of pre-stressed, reinforced polyester. One of Piano + Rogers' lesser-known projects was offices for the furniture company B&B Italia in Como, Italy (1972–3), a box suspended from a tubular portal frame that anticipates the **Sainsbury Centre** and Patscentre International in New Jersey. The Renzo Piano Building Workshop was established in 1981, its name hinting at a culture of craft production that has since been stretched to increasingly complex and far-flung projects.

In *Architecture and the 'Special Relationship'*, Murray Fraser and Joe Kerr describe high-tech as an Anglo-American cultural exchange, but architectural culture in the United States could be characterized as *Deutschamerikanisch*.[34] Born in Nuremberg, Helmut Jahn imbibed the

Indiana Bell Switching Center, Columbus, Indiana, 1978 by Caudill Rowlett Scott.

influence of Mies van der Rohe at the Illinois Institute of Technology and C F Murphy Associates of Chicago, where he became director of design from 1973. At the Kemper Arena, Kansas City (1973–4), the roof deck is hung from three external space trusses. A turn to slick skins is seen at the St Mary's Athletic Facility, Notre Dame, Indiana (1976–7), and the De La Garza Career Center, Chicago (1980). The State of Illinois Center, Chicago (1980–5, renamed the James R Thompson Center), was all atrium, decked out in pulsating salmon, silver and blue, the architectural equivalent of Chicago House music. Built to house state employees, it has been bought up and stripped out by Google.

The lineage of the Case Study Houses was continued by Helmut C Schulitz, who developed the TEST system (Team for Experimental Systems and building Techniques) with a small module that admitted a wide range of standardized components. As a prototype, he designed a family residence for himself, cantilevered over a steeply sloping site in Beverly Hills, California (completed 1976). In the pair of houses for his family and Roger Simon's in Laurel Canyon nearby, Peter de Bretteville took the similarly Eamesian approach of choosing and articulating standard catalogue items to a create 'a strongly constructed object, bolted, trussed and braced'.[35]

Technological innovation might be expected at Columbus, Indiana, the Midwestern home of the Cummins Engine Company. A foundation established by its CEO, J Irwin Miller (1909–2004), doled out civic buildings to world-class architects. At Hardy Holzman Pfeiffer Associates' Mount Healthy Elementary School (1970–2) and Columbus Occupational Health Association (1973–4) brightly coloured ducts and trusses play up elaborate open-plan layouts with rotated grids and counter axes. A pop sensibility is also on display at the Fodrea Community School (1973) and Indiana Bell Switching Center (1978) by Paul A Kennon, successively design principal and president of Caudill Rowlett Scott (CRS).

Kennon also oversaw the Thomas A Leavy Activities Center at the University of Santa Clara (1975), then the world's largest air-supported structure; the Herman Miller Seating Plant in Holland, Michigan (1980), which parallels the work of Nicholas Grimshaw & Partners, including their **Herman Miller Factory**; and the Carver-Hawkeye Arena at the University of Iowa (1980–2), with its external 'Takenaka' space trusses of weathering steel. The San Franciscan practice Holt Hinshaw Pfau Jones developed an industrial noir aesthetic with their kinetic Astronauts' Memorial at the Kennedy Space Center, Cape Canaveral (1989–90); the film and tape archive for Paramount Pictures in Hollywood (1989–91); and the sinister mechanisms of the UCLA Chiller-Cogeneration Plant (1994).

In Japan, the technological impulses of the Metabolist movement and Expo '70 were subsumed into divergent aesthetics. The work of Itsuko Hasegawa, Toyo Itō, Takefumi Aida, Hiroshi Hara, Tadao Ando, Hiromi Fujii and Kazunari Sakamoto, all born between 1933 and 1943, was exhibited in 1978 as 'A New Wave of Japanese Architecture'.[36] Shin Takamatsu became known for monstrous, mechanistic toys, alternatively ritualistic or humorous, while Riken Yamamoto designed urban townhouses for extended families in which interlocking volumes and terraces are swept by curved, Teflon-clad canopies. Shelter is equally pertinent to Toyo Itō's Silver Hut, Tokyo (1982–4), which combines the prototypes of primitive hut and spaceship. In the early work of Itsuko Hasegawa, light is modulated through perforated aluminium screens, glass blocks or cupolas set in pyramidal roofs. On an urban scale is the work of Hiroshi Hara, including the Umeda Sky Building in Osaka (completed 1993) and Kyoto Station (completed 1997).

ASPECTS OF TECHNOLOGY

Steel versus concrete would be one way to cast the story of constructional innovation in the 20th century. The ubiquity of reinforced concrete was a product of Britain's straitened post-war economy, in which steel was pricey and hard to obtain. Concrete was made to work harder, and technical developments included skinny shell structures, precast and pre-stressed concrete, folded slabs and slipform construction. As the balance between labour and capital shifted, energy-hungry steel became more viable, starting with single-storey construction. But it was more than a question of economics. Reliance Controls marks the first time that Team 4 and their engineer, Tony Hunt, had worked in exposed steel, and they looked west towards the American steel-building tradition. By the time of the Centre Pompidou steel had once more become a vehicle for constructional innovation.

In particular, Pompidou showed off the untapped potential of tubular steel, which soon became synonymous with the high-tech aesthetic. Circular or square hollow sections (SHS) were hardly new – the Forth Bridge is made from them – but it was only after mid-century advances in preparation, welding and connections that their potential could be exploited. Aside from their visual appeal, tubes make a good structural form: they are efficient, work well in compression, form curves with ease, resist bending and torsion and their low surface area helps with corrosion and fire protection. They can be used as conduits for services or filled with concrete or water for fire resistance.

Behind much technological innovation in architecture lay the expert input of specialist subcontractors. A signal example is Tubeworkers of Claverdon, Warwickshire, who more than any other British firm helped to establish tubular steel in the high-tech vocabulary. They started out making standard structures but their managing director, John Hurst, came to realize that 'Pompidou was no flash in the pan' and that the smart money was on architectural one-offs.[37] After collaboration with the engineer Z S Makowski, the break-through came with the space-frame roof at **Willis Faber & Dumas** and Piano + Rogers' UOP factory at Tadworth, Surrey. Contracts for the **Sainsbury Centre**, **Inmos**, the **Renault Distribution Centre**, **Lloyd's**, **Schlumberger** and **Stansted** followed. Hurst gained repeat business because he and his engineering director Keith Ford were prepared to sit down and talk designers through the steelwork details. As Martin Manning of Arup explained, 'John did not take advantage when we made mistakes. He was a contractor that engineers and architects – including Foster, Rogers and Grimshaw – enjoyed working with because of his abilities.'[38]

Suspended structures are one of the leitmotifs of high-tech. Tension in architecture has a long history, stretching back to tents and cable-stayed bridges via trailblazing pavilions, terminals, hangars, service stations and machine halls of the 19th and 20th centuries. Ove Arup & Partners fostered a brood of two-way masted structures, including Hall 7 at the Birmingham NEC (1979–80; Edward D Mills & Partners); the Fleetguard Factory at Quimper, Brittany (1979–81); and Foster's **Renault Distribution Centre**. Of greater sophistication were Frei Otto's experiments with pre-tensioned cable networks and fabric membranes. These technologies became so bound up with Buro Happold and Arup's that Tony Hunt dubbed them 'the membrane mafia'.[39]

In *Theory and Design in the First Machine Age*, Banham criticized modern architects for their failure to exploit the materials and techniques of their own age, such as alloys,

Page 29: Fleetguard Factory, Brittainy, France, 1979-80, Richard Rogers Partnership.

plastics, synthetic fabrics, laminated wood and lightweight monocoque enclosures. High-tech was a belated attempt to catch up with materials science, incorporating glass-reinforced plastic (sometimes known as fibreglass), cast steel, ductile iron, superplastic aluminium alloys, space frames, gridshell and cable-net roofs, diagrid façades, structural glazing, solar, low-iron and fritted glass, intumescent coatings, double-skin façades, photovoltaic cells, tensile fabric membranes coated in PVC or Teflon, inflated ETFE cushions, ultra high-performance concrete, passive chilled beams and ground source heat pumps. Not all of it worked. The **Sainsbury Centre** closed for several months in 1988 after the failure of its cladding panels, whose core of phenolic foam insulation started to react with the aluminium shell. Nickel sulphide inclusions in toughened glass were linked to the cracking or shattering of panes at Grimshaw's **Waterloo International Terminal** and **Western Morning News**.

Central to high-tech was the concept of 'technological transfer'. This was the appropriation of advanced technology – and its imagery – from other sectors, particularly the automotive, telecommunication and aerospace industries. The emblematic example is the neoprene or 'zipper' gasket used for sealing the windows of automobiles, used as early as 1948 by Carl Koch to join the panels of his prototype Acorn House and picked up by Team 4 and others in the mid-1960s.[40] But technology is a moving target: not long after Richard Horden deployed aluminium masts as columns in his **Yacht House**, yachts started sporting carbon-fibre masts. The analogy between a building and a vehicle or a machine eventually breaks down because of the different (and perhaps incompatible) way they are developed. As Peter Rice explained, 'cars and aeroplanes are not designed, they are improved'.[41] The fundamental difference, Jan Kaplický believed, was that whereas performance is the principal criterion for the success or failure of a car or a yacht or a helicopter, the construction industry is founded on real-estate values.[42]

High-tech architecture is assembled rather than built. Components are prefabricated at the factory and clipped into place on site. The key distinction is whether the parts are off-the-peg or tailor-made. Much early innovation in prefabrication took the form of 'closed' kits of parts, such as those of the post-war Hertfordshire schools and the CLASP system.[43] But what most architects wanted was the ability to combine commercially available products on an 'open' framework of common dimensions. This can be done in the spirit of the engineer or the *bricoleur*, to invoke Claude Lévi-Strauss's distinction.[44] Industrial components were artfully combined at the Eaves House and the Maison Jean Prouvé, improvised from leftover components from the designer's workshop.[45] The alternative, of bespoke components and assemblies drawn up from scratch, offered the attraction of total control, although usually at greater cost and risk. Tom Barker of Ove Arup & Partners was responsible for the building services at **Lloyd's**:

> *We found that products weren't on the market to satisfy all our requirements. [...] We had to get manufacturers to come along with us in developing the products. There was a lot of pressure because we were working in the context of a building contract. Once you move out of the catalogue you're living dangerously.*[46]

If these 'specials' were deemed to meet a wider need they might be added to a supplier's catalogue. Informal partnerships between architects and product manufacturers were symbiotic; the designer was a free source of research and development while taking advantage of surplus production capacity.

High-tech was suffused with a Ruskinian attachment to craft and the artisan. While its components looked mass-produced, more often they were made in small batches. The pitted, cast surface of the gerberettes at the Centre Pompidou manifested what Peter Rice described as '*traces de la main*', proclaiming that they were made by people, not machines. High-tech was capable of embracing a DIY ethic: the red, Meccano-like atrium roof of Newlands Primary School in Yateley, Hampshire (1979–80) was fabricated by the industrial designer Tony Pritchard in his garage and driven to site in his car boot. In the case of construction kits such as **Patera** and Ian Ritchie and Peter Rice's Shelterspan, entrepreneurial manufacturers were directly involved in the design process. Small-scale cooperation was, Peter Cook observed, characteristic of 'that stream of English fiddlers-with-cars, builders-of-boats, cannibalisers and people who always seem to have another way of using a working part from the one that you expect'.[47]

IS THIS TOMORROW?

What happened to high-tech and where, if anywhere, is it today? Did technology leave architects behind, as Reyner Banham feared, or was it the other way around? High-tech tended to be criticized from the left for its links to big business and from the right for its rejection of tradition and context. It was around the time that the Prince of Wales first took aim at modern architecture that high-tech architects started to engage with urbanism and culture. The challenge of adding to **Lord's Cricket Ground** and **Bracken House** prompted Michael Hopkins to examine historic materials and precedents, an approach he summed up as 'technology comes to town'.[48] Foster Associates cultivated a sense of place at Carré d'Art Nîmes (on one sketch Foster scribbled 'no diagonals in structure – must not look "industrial"') and an unrealized project for BBC headquarters at Portland Place.[49]

High-tech's contextual mode was sufficiently polite for institutional and infrastructure commissions, and in the 1990s became the accepted identity of a wave of lottery-funded, millennial projects. By this time, the 'big four' practices had been taken on by major clients in Europe and the Far East. The primary colours, masts and pipes of high-tech's heroic period were shed like lost Lego parts, leaving a corporate Esperanto of white tubular steel and green glass. Looking back, Owen Hatherley regretted that the weird, exciting and sometimes sinister early works had given way to an international mainstream, mirroring what had happened to modernism.[50]

The picture changes slightly if high-tech is regarded as part of a wider set of attitudes to technology rather than a style with a shelf life. If its underlying tactics – including architect-engineer dialogue, an engagement with the making process, and technological resourcefulness – have any present-day applications, they may be in the field of sustainable architecture. This may seem paradoxical, given that many high-tech buildings were highly dependent on fossil fuels in their manufacture and operation (although this book also profiles early instances of energy efficiency, renewables and sustainable building materials). If the radical and disruptive spirit of the high-tech pioneers can be found anywhere today, it is in the handful of engineers and architects who are rethinking E F Schumacher's principles of appropriate technology and 'small is beautiful' to make more efficient, locally sourced and resilient buildings. This includes engineers such as Steve Webb and architects like Gilles Perraudin who are finding new structural roles for low-carbon materials like structural stone and cross-laminated timber (CLT).

Locally woven bamboo courtyard canopy for the Third Space community centre in Udaipur, India, designed by Studio Saar and Webb Yates.

New buildings are only part of the answer. The preservation, conservation and retrofitting of architecture of the recent past – not least high-tech heritage – demands just as forward-thinking an approach. Listing is often ahead of the curve of appreciation and preservation. In 1991, just 16 years after it was put up, **Willis Faber & Dumas** became the first post-war building to be listed at grade I. In the following years, the **Snowdon Aviary**, **Olivetti Training Centre** and **Cummins Engine Assembly Plant** were listed at high grades, following threats of demolition or alteration. There were also several high-profile losses, such as Reliance Controls, demolished in 1991.

Since then, the discovery of high-tech architecture by younger generations has coincided with the passing of some of its pioneers. There have been major retrospectives of Richard Rogers (2013, Royal Academy) and Norman Foster (2023, Centre Pompidou), while the story of their generation was related by

The Brits Who Built the Modern World, a 2014 BBC TV series paired with a RIBA exhibition. High-tech was the subject of 'Superstructures: the New Architecture 1960–90', a 2018 exhibition at the **Sainsbury Centre**.

Scholarly reassessments have included Todd Gannon's *Reyner Banham and the Paradoxes of High Tech* (2017) and Angus Macdonald's *High Tech Architecture: A Style Reconsidered* (2019), a critical reappraisal of the movement from a structural engineer's perspective. The relationship between Archigram and British high-tech is the subject of Annette Fierro's *Architectures of the Technopolis* (2023), while *High-Tech Heritage: (Im)permanence of Innovative Architecture* (2024) considers the conservation challenges of technologically innovative architecture.

This book reassesses the built legacy of British high-tech half a century on from its emergence. Some of the pioneers are long gone or barely recognizable, while others have been lovingly stripped down and renewed like a classic motor or a Japanese temple. An architecture that accepts the prospect of change and impermanence might place less value on the capacity to age gracefully with a time-worn patina – a quality that is sometimes fetishized. This can present philosophical and practical dilemmas when it comes to questions of preservation and conservation. Nevertheless, there is mounting opposition to the needless demolition of buildings erected within our lifetimes – especially those, like high-tech, with high levels of embodied carbon. With it comes a growing desire to protect and enjoy the best examples of this extraordinary moment in architecture.

Pages 34–5: Patera prototype building, Albert Island, Royal Docks, London, 1981–2, Michael Hopkins & Partners.

A NOTE ON THE BUILDING ENTRIES – AND ON LISTING

The entries in this book span the period 1962 to 2019 and are arranged chronologically by building type. Most are habitable buildings; engineering structures such as transmitter masts, pylons, cooling towers, dams, radio telescopes and wind turbines fall outside its scope. Also excluded are bridges, viaducts and walkways – those designed by high-tech architects warrant a full-length study of their own. As this is a book about buildings, dates in the subheadings refer to the construction period. A building name set in **bold type** indicates that it has an entry of its own. Underlined terms denote an entry in the glossary at the back of the book.

Almost a quarter of the entries in this book relate to listed buildings, although the proportion is likely to increase over time. Their entries include the date and grade of the listing. Buildings which are of 'special architectural or historic interest' can be listed, which gives them legal protection. England and Wales has a system of three grades, where grade I denotes exceptional special interest and grade II* more than special interest. About 90 per cent of all listed buildings are grade II, which are of special interest. Scotland has a separate system; category A buildings are of national or international architectural or historic importance, category B are significant at a regional level or hold more than local importance, and category C buildings hold local importance.

Under the 'thirty year rule' introduced in England and Wales in 1987, any building over 30 years can be considered for listing. Younger buildings are normally listed only if they are of outstanding quality (generally interpreted as being equivalent to grade I or II*). When buildings are assessed for listing in England and Wales, the age of a building is calculated from the point at which the ground was first broken.

residential

125 PARK ROAD

St John's Wood, London
1968–70; Farrell/Grimshaw Partnership
Structural engineer: Anthony Hunt Associates
Listed at grade II in 2001

Housing associations grew rapidly in the 1970s, providing a third way besides owner occupation and council housing. In 1966 Terry Farrell and Nick Grimshaw became founder members of the Mercury Housing Society Ltd, a co-ownership society that obtained a third of its finances from the government's Housing Corporation and the remainder on a 40-year mortgage from the Cheltenham and Gloucester Building Society. On a site overlooking Regent's Park they planned an 11-storey block of 41 flats. The small site suggested a square plan with a generous central core to take wind loads, services and circulation. Around the perimeter ran freestanding columns, heating and electrical sockets, and continuous glazing strips. The compact layout accommodated two one-bedroom and two two-bedroom flats per floor, with four duplex penthouses and a caretaker's flat, all built to Parker Morris space standards.[51] Movable partitions and fitted furniture were precluded by the stringent cost limits, so the interiors were left as serviced shells that owners could fit out to their taste.

Rounded corner windows allowed panoramic views. Corrugated anodized aluminium was chosen as the cladding material – curved corner sheets cost sixpence extra. Park Road shows Grimshaw's early interest in lightweight, industrial materials, while reflecting shifts in patronage from council housing towards a voluntary model. With others from Mercury, Farrell went on to set up the Maunsel Housing Society, for whom he designed several colourful, low-cost schemes.

Further reading:

Terry Farrell, 2004, *Place: A Story of Modelmaking, Menageries and Paper Rounds*. London: Laurence King Publishing.

Pages 36–7: Hopkins House, Hampstead, London, 1975–6, Michael and Patty Hopkins.

PARK ROAD NW8

SPENDER HOUSE

The Studio, Ulting, Maldon, Essex
1968–9; Richard + Su Rogers with John Young
Structural engineer: Anthony Hunt Associates
Listed at grade II in 2012

The artist and photojournalist Humphrey Spender (1910–2005) is best known for his work for *Left Review*, *Picture Post* and the Mass Observation Project, including a remarkable study of Worktown (Bolton).[52] With his wife, Pauline, he wanted a new house and studio in the orchards of their Victorian rectory. A student at the Royal College of Art, where Spender taught textile design, put him in touch with Richard and Su Rogers. Completed five years later, the Spender House was their first project after the break-up of Team 4 (pages 15–16).

As at the Eames House, residence and studio form twin pavilions enclosing an outdoor room. Bolted portal frames are placed outside the walls, cross-braced in the short dimension. They span lengthways, permitting the long walls to be fully glazed in nylon-coated aluminium frames; the remaining walls and roofs are clad in PVC-coated corrugated steel. The studio is windowless to aid concentration; illumination comes from a peaked skylight. Bright accents are provided by the chrome yellow steelwork and purple sliding doors. Like Charles and Ray Eames, Spender quietly subverted the open plan by filling it with Victoriana, eclectic furnishings and *objets trouvés*. The house proved to be a one-off rather than the prototype its designers hoped it would become, and their client complained of condensation and heat loss. Nevertheless, it remains an important milestone, translating to the domestic sphere the interest of Richard and Su Rogers in industrial materials and prefabrication techniques.

ROGERS HOUSE

22 Parkside, Wimbledon, London
1969–70; Richard + Su Rogers with John Young
Structural engineer: Anthony Hunt Associates
Listed at grade II* in 2013

With the **Spender House**, the Rogers House forms a link that joins Reliance Controls (1965) to the Zip-Up House project of 1968. It was designed for Richard Rogers' parents Nino and Dada, on a garden site near Wimbledon Common. The layout, adapted from the Spender House, comprised two pavilions – the main residence and a studio with attached car port – enclosing a paved courtyard. A row of welded portal frames, external in the Spender House, was brought inside to avoid cold bridging. The result is a tube with glazed ends. For the side walls John Young found an insulated panel, developed by Alcoa for refrigerated trucks. They were made of PVC-laminated aluminium, sealed with a 'snug seam' neoprene joint. Into the cladding were punched curved-corner openings obtained from a manufacturer of bus windows.

The 12.2m (40ft) clear spans set up an open-plan interior, punctuated by the yellow-painted flanges of the frame and dominated by a freestanding kitchen counter, like that at Creekvean (1964–7, Team 4). 'My mother loved to cook,' Rogers recalled, 'so the open plan kitchen is the heart of the house where family and friends could gather, very much in the Italian way.'[53] The house is detailed with flair and finished with a pop palette of chrome yellow and acid green. Sliding partitions can be drawn across the bedrooms and study behind. In 2015 Richard Rogers gifted the house to the Harvard Graduate School of Design to host visiting fellows. The building was later overhauled by Rogers alumnus Philip Gumuchdjian, who renewed the skin and services, removed later additions and upgraded its energy performance.

PILLWOOD

Pill Creek, Feock, Truro, Cornwall
1973–4; Colquhoun and Miller
Structural engineer: Anthony Hunt Associates
Listed at grade II* in 2017

While Team 4's Creekvean (1964–7) is a masonry structure conceived by high-tech pioneers, its neighbour across Pill Creek is a house of glass, steel and plastic by a practice known for its heavyweight, tectonic approach. Both buildings were designed for Marcus and Rene Brumwell by their daughter Su Rogers. In 1971 the Brumwells looked to build a 'holiday perch', financed by the sale of a Mondrian. Su had recently parted from Richard Rogers and had got together with John Miller. Sited on a secluded wooded hillside, the house is built against a retaining wall with steps up to a rear terrace.

Pillwood is a geometrical game of open and closed: white GRP panels dissolve into a cascading roof of patent glazing that follows the fall of the land. A flexible layout of sliding partitions was devised to accommodate groups of different sizes and generations. A first-floor dining room and bedrooms below open onto the double-height conservatory, connected by spiral staircases at the front and rear. The triangular conservatory incorporates sail-like blinds and louvres to shade the sun, yet the design patently predates the oil crisis. Like the Melrose Centre in Milton Keynes, Pillwood is the progeny of Miller's 'love affair with industrialized building materials'.[54] The house invites comparison with the pure, hard geometry of Stirling and Gowan's work, yet Gowan compared it to Albert Frey's Canvas Weekend House at Long Island (1933–4), in which drapes subdivide an open plan.

HOPKINS HOUSE

49a Downshire Hill, Hampstead, London
1975–6; Michael and Patty Hopkins
Structural engineer: Anthony Hunt Associates
Listed at grade II* in 2018

Michael and Patty Hopkins were among the original occupants of Neave Brown's narrow-fronted terrace at Winscombe Street in Archway (1965–6). When their family outgrew it they decided to build a very different house. As well as two architects and three children the residence also accommodated their fledgling office – only in 1975 did Michael Hopkins strike out from partnership with Norman Foster to establish a practice with Patty. Their former boss Frederick Gibberd became their next-door neighbour after he sold them a plot parcelled off from his own garden. Their engineer was Tony Hunt, another former Winscombe Street resident.

Materials and techniques learned by the Hopkins from industrial projects were translated to a domestic scale. With the Eames House and **IBM Cosham** (on which Michael Hopkins was partner in charge) is shared a pragmatic yet refined approach to steel-framed construction based on small spans, slender columns, cross-bracing and open-web beams. Side walls are of profiled steel sheet, while front and back are fully glazed and screened by venetian blinds. Due to the 3m (10ft) drop between pavement and garden, the two-storey house presents a single-storey façade to the street, from which it is entered via a metal footbridge. Zones for working, sleeping and sitting were provided at street level, while the garden floor accommodated three children's bedrooms and kitchen-dining space, all arranged around an off-centre bathroom core and a tubular steel spiral staircase. Partitioning was initially limited to venetian blinds and full-height bookcases, but the plan subsequently evolved as first the practice then the children grew up and moved out.

THE CRYSTAL PALACE

Friar's Lane, Bury St Edmunds, Suffolk
1977–8; initial design by Michael and Patty Hopkins
Structural engineer: Anthony Hunt Associates

Flexibility is a key precept of high-tech architecture, although most designers regard change as something that occurs within a preconceived framework. From the outset, the Crystal Palace evolved in a more organic and unpredictable manner, with the result that its architects have not admitted it to their oeuvre. It is all the more interesting for that. The house was built for Peter Tracey, an inventor and sailor, who met the Michael and Patty when he sold them a gaff-rigged cruiser. Tracey acquired a fine site, a south-facing field running down to the River Linnet and water meadows beyond. This became the setting for a pavilion of steel and glass, raised on low stilts. Details such as the open-web beams, cross-bracing and ribbed roof deck recall the **Hopkins House**. More reminiscent of Australian steel-framed houses is the veranda, which extends around three sides. To the north, a carport and covered entrance complete a square plan. The layout is like a miniature Roman villa, with interconnecting rooms around a central atrium.

After architect and client parted ways, construction was overseen by Tracey, who departed from certain Hopkins details while installing features of his own, such as the sliding window frames, heating system and an idiosyncratic wet room with a canvas bath slung on a tubular frame. Further alterations were made by the next owner, including glazing over the central courtyard and enclosing part of the carport. In 2014–15 the house was renovated by James Soane of Project Orange, who overhauled the services and opened up the interior.

SPENCE AND WEBSTER HOUSES

2c and 2d Belsize Park Gardens, Camden, London
1979–81; Spence and Webster
Structural engineer: John Carter

These single-storey houses were designed by Robin Spence and Robin Webster for themselves and their families as a mirrored pair sharing a central court. They obtained the site – the garden of the neighbouring stuccoed house – for a reasonable sum as several previous development proposals had been turned down. Theirs was better received by the Camden planning officer (Duncan Macintosh), who happened to have written a book on courtyard houses.[55] The roof is a profiled metal deck in three spans of 6m (20ft), supported by hollow section beams. Instead of load-bearing partitions or internal columns, the roof is held up by skinny perimeter columns on 1.8m (6ft) centres. This forms the module for the aluminium double-glazed windows, with sliding units front and rear. Unlike the **Hopkins House** there are no cross-braces; the roof runs across the two houses keeping the structure rigid.

The plan is divided into three zones. To the rear, a living area and master bedroom overlook the garden, while a family space adjoins the courtyard entrance, with the kitchen, utility room and bathrooms forming a service core behind. The front section could be partitioned for children's bedrooms, a self-contained flat or an office. Beneath the central zone is a basement that forms part of the concrete foundations. In the 1960s Spence worked for Myron Goldsmith of SOM's Chicago office, an experience that informed his subsequent steel-framed houses in Thornton, Leicester (1968), and Hunstanton, Norfolk (1971). Here the influence is the featherweight framing and open plans of the West Coast Case Study House programme.

MCMANUS HOUSE

Longfield, New Road, Princes Risborough, Buckinghamshire
1980–2; Aldington, Craig and Collinge
Structural engineer: Anthony Hunt Associates

Aldington, Craig and Collinge are known for beautifully put-together houses of brick and timber, carefully integrated with the landscape. A regard for making is no less evident at the McManus House, a collaboration with Tony Hunt in which the glazing assembly was designed and fabricated by Tony Pritchard, the industrial designer who acted as cladding consultant on the **Sainsbury Centre**. A house of metal and glass suited their client, John McManus, who ran a car bodywork firm in Wembley. A single-storey pavilion is hidden behind an inter-war bungalow, joined by a link passage (which accommodated McManus's book collection).

Open-plan living and dining spaces are punctuated by a kitchen core on axis with the passage, while a small swimming pool is fitted into the adjacent courtyard. Portal frames are internally exposed, as are the tension wires that cross-brace the roof and the underside of the profiled aluminium roof deck. Full-height doubled-glazed window units are sealed with neoprene gaskets into extruded aluminium frames with curved corners of cast aluminium. The end walls are clad in corrugated GRP panels, a material also used for the wide, moulded doors set on pivot hinges. A car port with castellated beams was added beside the bungalow. An affinity for the materials, colours and tolerances of high-tech is also manifest in the practice's Mechanized Letter Office at Hemel Hempstead (1983–5, demolished) and Paul Collinge's conversion of Lemsford Mill near Welwyn Garden City (1981–3).

Further reading:

Peter Aldington, 2016, *Houses: Created by Peter Aldington*. London: RIBA Publishing.

EAGLE ROCK

London Road, Uckfield, East Sussex
1981–2; Ian Ritchie Architects
Structural engineer: Anthony Hunt Associates

Eagle Rock seems closer to a chassis or a site-specific installation than the received image of a steel-framed house. It was built for Ursula Colahan, a botanist and the mother-in-law of Martin Francis, Ritchie's friend and future practice partner. In discussing her return to England from Italy, she arrived at the simile of a bird on the wing, carrying a suitcase. From the pages of *Exchange & Mart* came a site, a secluded clearing in woodland, sheltered by sandstone outcrops. At first Ritchie thought of a base camp, a set of wooden containers among the rocks, over which a transparent canopy could be draped. This evolved into an 'anchored sculpture' that returned to Colahan's image of flight and freedom, of an eagle on the rocks or a bow and arrow.

A pair of swept-wing canopies (accommodating bedrooms and lounge) are suspended from a fuselage (dining, kitchen, utility) with a detached greenhouse and carport fore and aft. The canopies and the central spine are suspended from a stout tubular A-frame, anticipating PA Technology Laboratories in Princeton, New Jersey (1982–5, Richard Rogers Partnership). Within the open cage nestles an elevated plant room: the notional head. External louvres were installed, rippling like ruffled feathers (an idea that recurs in the kinetic south façade of Ritchie's Sainsbury Wellcome Centre of 2016). The detailing of the windows and wall panels is restrained and pragmatic; the emphasis is firmly on the idiosyncratic gesture of the perched structure.

Further reading:

Ian Ritchie, 2013, *Being: an Architect, volume 1*. London: Royal Academy of Arts.

YACHT HOUSE

Windrush, The Common, Woodgreen, Fordingbridge, Hampshire
1984; Richard Horden Associates
Structural engineer: Felix J Samuely & Partners

Architects, Richard Horden felt, have a sibling relationship with the builders of automobiles, boats, aeroplanes and spacecraft. A keen sailor, he collaborated with Ian Proctor, a mast manufacturer, on the design of a flat-pack construction kit. This house, built for his sister in the New Forest, served as the prototype. It is based on the technology of slimline, lightweight, demountable yacht components. Aluminium masts serve as columns, oval-section spars as beams, rigging as tension braces, deck hatches as rooflights and blue sail awnings as canopies. These components are brought together on a square grid (or 'wind frame') of 20 bays that weigh 2.6kg/m^2 (0.53lb/ft^2). The house was assembled at weekends by Horden, his client and other family members.

The result is lighter and more informal than the Miesian house he built for his parents at Poole in 1973–5. Like Craig Ellwood's Case Study House 18, some of the 3.6m (12ft) bays are left open as patios and courts, lightly screened with square trellises or external aluminium louvres (the latter recalling the **Sainsbury Centre** on which Horden worked). The gestures towards open-endedness were not carried into the interior which, at the client's request, was finished with plastered stud walls. In a similar vein, the roof deck was envisaged as an Eamesian patchwork of panels, canopies and roof lights, kitted out with nautical paraphernalia: a satellite dish, solar panels, anemometer and flagpole. Horden's system was refined at several later projects, including the SkiHaus (1992), a portable alpine hut.

Further reading:

Richard Horden, 1995, *Light-Tech: Towards a Light Architecture*. Basel: Birkhäuser.

GRAND UNION WALK

Kentish Town Road, Camden, London
1986–8; Nicholas Grimshaw & Partners
Structural engineer: Kenchington Little & Partners
Listed at grade II in 2019

Two decades after **125 Park Road**, a mixed-use, inner-city development including a **Sainsbury's** supermarket allowed Grimshaw to return to urban housing. Grand Union Walk is located along the former Kentish Town Wharf of the Grand Union Canal, constructed in 1812–16 as the Regent's Canal. At Grimshaw's suggestion an existing scheme for a block of flats was switched to a terrace of freehold houses. There are ten three-bedroom dwellings (one of which was acquired by Grimshaw himself), with a one-bedroom maisonette and a bedsit flat at the end. Front doors are reached from a residents' path along the canal, in turn accessed from an entrance on the Kentish Town Road bridge. The wide frontages compensate for the shallow plots and the absence of south-facing windows (the housing backs onto the supermarket's loading bays).

The houses are planned across three floors. The ground floor is raised above the towpath for privacy and contains the entrance hall and a bedroom-cum-study. Above is an open-plan kitchen-diner with a top-lit, double-height space overlooking the canal. An aluminium window-wall can be electronically raised to throw the dining area open to the first-floor balcony. Construction is relatively conventional: precast concrete joists bearing onto cross walls of concrete blockwork. Grimshaw opted for deliberately industrial imagery to reflect the canalside setting. Each house has two bays: a narrow, recessed, glazed one alternates with a wide, aluminium-skinned capsule with curved-corner neoprene gasketed windows. In a gesture worthy of Bucky Fuller, each occupant was presented with a ring-bound user's manual.

HAUER-KING HOUSE

40 Douglas Road, Canonbury, Islington, London
1993–4; Future Systems
Structural engineer: Anthony Hunt Associates

This family house, designed by Jan Kaplický and Amanda Levete for their friends Debra Hauer and Jeremy King, provides a fresh take on an old model: the narrow-frontage townhouse. Brick cross walls are spanned by a glass enclosure that frames the mature trees and changing seasons outside. The entrance front is robust and defensive. A drawbridge-like ramp skirts around a weeping ash (since felled) towards the front door, set in a 7m-tall (23ft) glass screen of glass blocks. Above, a cascade of bolt-fixed glazing sweeps up and over in cranked sections, returning at 50 degrees as an inclined, south-east-facing rear wall. The floors are pulled back to form a full-height entrance lobby and, at the rear, a mezzanine living room over a basement kitchen-diner.

There is a sense that detailing provided client and architect alike with a respite from the struggle of putting up a house. Lacquered bathroom pods are finished in egg-yolk yellow while the kitchen's island unit is a sculptural object. The stair is a composition of looped tubular handrails and extruded aluminium treads cantilevered from a central spine. It is a tight-fit design in which little is left to chance – except, perhaps, the building's environmental performance. The machine aesthetic looks the part yet, as David Jenkins observed, 'what one is dealing with here is not a production model but an exquisitely hand-built prototype'.[56]

Further reading:

Martin Pawley, 1997, *Hauer-King House: Future Systems*. London: Phaidon.

MALATOR

Druidston, Broad Haven, Haverfordwest, Pembrokeshire
1998; Future Systems
Structural engineer: Techniker

Despite its ancient origins, the earth-sheltered house remains relevant in sensitive or exposed locations, or where thermal or noise insulation is paramount. Arthur Quarmby designed several, including Underhill (1973–5, grade II), his home in the Peak District National Park. On a large scale was Archigram's winning entry for a subterranean entertainment building at Monte Carlo, with engineering by Frank Newby. Predictably, Malator became known by locals as 'the Tellytubby house', after the green-domed abode of the BBC children's TV characters. It was designed as a weekend home for Gill Marshall-Andrews and her husband Bob, then Labour MP for Medway. Their site overlooked St Brides Bay in the Pembrokeshire National Park, so to minimize its visual impact a pair of embankments were formed around the curved structure, which was topped with an 80mm-thick (3in) layer of turf. The earthworks are scooped away front and back to reveal glass walls with circular opening lights. The architects likened the seaward elevation to an eye keeping watch over the Atlantic.

The plan is shaped like a tangerine segment. At its heart is an open-plan living space, with an island seating unit gathered around a wood-burning stove. It is enclosed by a pair of teardrop-shaped bathroom pods, acid yellow in colour. Into the rounded corners go the master bedroom and a pair of guest bedrooms. Construction methods were kept relatively straightforward at such a remote site, the roof sitting on a retaining wall and a continuous steel ring beam.

Further reading:

Ivan Margolius, 2022, *Jan Kaplický – For the Future and for Beauty*. Fellbach: Edition Axel Menges.

2
public

FARNBOROUGH COLLEGE OF TECHNOLOGY

Boundary Road, Farnborough, Hampshire
1984–6; Hampshire County Council architect's department
Structural engineer: Anthony Hunt Associates

Under the leadership of county architect Colin Stansfield Smith, Hampshire became a beacon of public sector architecture in the 1980s. Although his team gained a reputation for warm brick, stained joinery and big roofs, there were also forays into high-tech vocabulary, with the help of Tony Hunt. A kit-of-parts approach seemed appropriate at Farnborough Tech, which started life in 1918 as the 'trade lads' school' of the nearby Royal Aircraft Establishment. A major programme of expansion provided the opportunity to consolidate a scatter of 1950s pavilions.

Faced with a crash programme and a sketchy brief, project architects Mervyn Perkins and Peter Galloway decided to provide an overarching umbrella of structure and services under which a flexible educational programme could take shape. As at Crestwood Secondary School (1981–2) and the John Darling Mall (1985; both at Eastleigh) a spine plan is organized around a glazed, full-height arcade or 'mall' – the retail terminology hinting at the marketization of education. Daylight is drawn into the cellular interior through triangular rooflights on structural trees and the mall's high barrel vault, whose tunnel-like form evokes the monumental portable airship hangar a short walk away. The unheated main concourse was intended to function as a noise and heat buffer, protecting the classrooms from aircraft noise and preheating the air supply. Recent additions to the campus include the University Centre (2016, Nicholas Hare Architects) and the Aerospace Research and Innovation Centre (2020–1, Bond Bryan).

Further reading:

Richard Weston, 1991, *Schools of Thought: Hampshire Architecture 1974–1991*. Winchester: Hampshire County Council.

Pages 64–5: Prototype workshop, Hooke Park, Dorset, 1988–90, Ahrends Burton & Koralek with Frei Otto.

PROTOTYPE HOUSE AND WORKSHOP

Hooke Park, Hooke, Beaminster, Dorset
1985–6 (prototype house), 1988–90 (workshop); Ahrends Burton and Koralek with Frei Otto
Structural engineer: Buro Happold

Hooke Park was acquired in 1982 by the furniture maker John Makepeace, who taught joinery from Parnham House nearby. The 350 acres of woodland became the base of his Parnham Trust, an educational charity concerned with the sustainable use of timber. Makepeace planned a forest campus to train students in the manufacture of furniture and handicrafts from thinnings, the trees felled to promote growth of a managed forest. He appointed Frei Otto, an international authority in tensile structures, who in turn brought in Richard Burton and Ted Happold. Using sketches, models and computer analysis, they applied advanced engineering to 'green' (unseasoned) timber. Trunks were kept in the round to preserve their natural strength; their outer fibres are longitudinally pre-stressed, enabling them to stay in tension under wind loads. Creating an effective tension joint was crucial. Otto suggested creating composite beams by bolting or binding members together. More efficient was Happold's technique of using epoxy resin to glue steel bars into sockets at the ends of the timber.

At the prototype house, a catenary roof is supported on spruce thinnings held in tension between ridge and eaves cables. The humped workshop building comprises three linked shells, braced by a PVC membrane roof. It resembles the offspring of a tithe barn and an airship hangar. In 2002 Hooke Park was acquired by the Architectural Association School of Architecture, becoming their 'design and make' campus. Each addition to the site is treated as a group research project, perpetuating the experimental ethos of the first buildings.

FLEET INFANT SCHOOL

Velmead Road, Fleet, Hampshire
1985–6; Michael Hopkins & Partners
Engineer: Buro Happold
Listed at grade II in 2017

Colin Stansfield Smith gingered up his Hampshire team by farming out a share of the annual programme to practices including Ahrends Burton and Koralek, Edward Cullinan and Aldington, Craig and Collinge. When he commissioned Hopkins in 1984, Stansfield Smith had in mind the masted membrane structures the practice developed with Buro Happold for an unbuilt project for Basildon town square. Working with Happold's firm, Patty Hopkins developed a scheme in which a tent billowed out over a glazed envelope. Despite Stansfield Smith's support the scheme was opposed by education officers and rejected by the education committee.

A modified design substituted a low-pitched metal roof, with continuous ridge glazing toplighting an internal 'street'. Nine classrooms are equipped with quiet pods, rubber-floored practical areas and double doors opening onto a paved terrace. Hollow-section columns support cranked rafters and a roof of insulated metal decking. Ted Happold suggested that 'in concept the structure has its origin in those large French electricity pylons that support their wires on outriggers'.[57] The glazed walls are shaded by Teflon-coated PVC canopies, a vestige of the original scheme. Solar gain is mitigated by high-level louvres and ridge vents, which generate a stack effect. Fleet complements a high-tech strain at Hampshire that can be traced back to the sleek extensions to the system-built Frogmore Comprehensive School (1978–9, job architects Peter Galloway and Mervyn Perkins). In 1998, Hopkins returned to design a music pod on the north side of the school.

QUEEN'S INCLOSURE PRIMARY SCHOOL

Cornelius Drive, Cowplain, Waterlooville, Hampshire
1988–90; Hampshire County Council architect's department
Structural engineer: R J Watkinson & Partners

When job architect David Morriss's design for a timber-framed school came in over budget he was faced with a complete redesign a few days before the scheme came before the education committee. He turned to Hopkins' recently completed **Fleet Infant School**, introducing several small but significant tweaks. Influenced by the houses of the Australian architect Glenn Murcutt, Morriss opted for a low-slung, gull-wing profile made up from a pair of segmental vaults with a glazed barrel vault over the internal street. The class bases face north with woodland views. They are supplemented by activity 'pods' entered from the street rather than the classroom, allowing them to be used by older children for craft and cookery. Aluminium louvres with adjustable blades replace Hopkins' fabric awnings.

Morriss's silvery shed sits over a grassed meadow at the boundary with mature woodland, a remnant of the ancient Forest of Bere (hence 'Queen's Inclosure'). Half of the site is preserved as a meadow and an 'ecology paddock' occupies the north-east corner. The nine-class school was designed before the inception of the National Curriculum, but its plan was sufficiently flexible to allow the layout to be reorganized on the basis of subject areas rather than year groups.

QUEEN'S BUILDING, EMMANUEL COLLEGE

Emmanuel College, St Andrew's Street, Cambridge
1993–5, Michael Hopkins & Partners
Structural engineer: Buro Happold

Queen's Building is a freestanding auditorium of the same golden Ketton stone as Emmanuel's Wren chapel. It is pill-shaped, as if a pair of Sheldonian theatres had been set back to back. Sited in the forecourt of the Master's Lodge, it is aligned not with the adjacent courts but with Leonard Stokes' North Court across Emmanuel Street. Queen's Building provides performance space in the form of a raked and galleried auditorium accompanied by reception and seminar rooms. The entrance is discreetly located in a cross passage inside the continuous colonnade, while a freestanding stair tower of glass brick (recalling Hopkins' **Inland Revenue** offices) provides a fire exit.

Monolithic at first sight, the exterior is in fact a framework of piers, flat arches and infill panels, set as flush as a glass skin but with each element shaped or scored to show its structural role. As at Glyndebourne Opera House and **Portcullis House**, the verticals become successively slimmer with increasing height to indicate the diminishing loads they bear. The stone piers are pre-stressed by concealed post-tensioning rods of stainless steel. Tapering kneeler blocks support the floor slabs and incorporate stainless-steel nodes that provide access to the rods. One might mistake the ashlar for a surface veneer; only in the colonnade is the heft of the structure visible. The performance space is given away only by the blind panels at the south-west end. Inside, oak trusses with stainless-steel ties support a heavy acoustic roof, clad in lead. Queen's Building was hailed as an essay in Arts and Crafts contextualism, yet its real significance lies in the revival of stone as a structural medium.

THAMES WATER TOWER

Holland Park Roundabout, Shepherd's Bush, London
1994; Damien O'Sullivan and Tania Doufa
with Brookes Stacey Randall Fursdon
Structural engineer: Atelier One

The Thames Water Tower is a superterranean monument to the commissioning of the 80km (50 mile) Thames Water ring main, the largest single water-supply infrastructure project undertaken in the UK. It is the outcome of an ideas competition won by Damien O'Sullivan and Tania Doufa, two students at the Royal College of Art. Their concept was translated into reality by Brookes Stacey Randall Fursdon, the practice founded by Alan Brookes and Michael Stacey.

The tower performs two functions. A steel core houses one of three 15m (50ft) surge columns that relieve pressure in the system. Integrated into the glass cladding is a barometer that indicates changes in atmospheric pressure with a cascade of dyed water, which is sprayed onto the inner face of the cylinder. A 3m (10ft) base of stainless steel houses the pumps and recirculation vessels. The main shaft is constructed of a 12mm (0.5in) skin of toughened glass, fixed with stainless-steel castings that support access decks at 2m (6.5ft) intervals. Crowning the tower is a curved solar array that powers the digital barometer and control panel. Illuminated at night, the tower was intended to create kinetic sculpture from a mundane piece of urban infrastructure. Erected as an emblem of investment in the newly privatized water industry, three decades on the tower is poorly maintained, its hydro-barometer long having ceased to function.

PORTCULLIS HOUSE

1 Parliament Street, Westminster, London
1998–2000; Michael Hopkins & Partners
Structural and services engineer: Ove Arup & Partners

A competition for a new parliamentary building on Bridge Street was won in 1972 by Robin Spence and Robin Webster before being dropped on financial grounds. Two decades later, the site, opposite the Houses of Parliament, was required for an interchange station on the Jubilee Line Extension. In 1989 Michael Hopkins & Partners was engaged to design **Westminster Station** and a new parliamentary building above as two discrete but integrated projects. Portcullis House is a singular building, designed for a projected lifespan of 120 years and connected to an Underground station and the Palace of Westminster. Its plan takes the form of a hollow doughnut with ranges of MPs' offices and conference rooms enclosing a glazed courtyard. The latter's diagrid barrel roof invites comparisons with Hugh Herland's medieval hammer-beam roof for Westminster Hall and Pier Luigi Nervi's 1930s aircraft hangars at Orvieto, Italy.

The project marked Hopkins' first major foray into energy efficiency, including a ground source heat pump and low-energy servicing. Working with Arup's, the design team developed a 'fat façade', in which air ducts concealed behind the perimeter walls converge into roof-top ventilation turrets. This was the functional pretext for an exercise in contextualism, the portcullis image embodying the paradox of a public building to which it is incredibly hard to gain access. Portcullis House marks high-tech's acceptance by the ruling class, inviting comparisons with the Victorian acceptance of the Perpendicular phase of Gothic as a 'national style'.

HAMPDEN GURNEY CHURCH OF ENGLAND PRIMARY SCHOOL

13 Nutford Place, Marylebone, London
2000–2; Building Design Partnership
Structural and services engineer: Building Design Partnership

The rebuilding of this Inner London primary was financed by an 'enabling development' in which two blocks of flats (also designed by BDP) were squeezed onto the site. To compensate for the loss of outdoor space, the lead architect Tony McGuirk proposed a vertical school or 'children's tower' with integrated play areas.[58] Although some London board schools incorporated rooftop playgrounds, the model was the stacked open decks of Jan Duiker's Openluchtschool in Amsterdam (1929–30), which sits within a residential court.

Hampden Gurney School has a symmetrical layout that contrasts secluded study and extrovert play. An arrow-shaped classroom wing faces north, overlooking the courtyard. The street corner is dominated by the sweeping curve of the open play decks, which are connected to the teaching spaces by a bridge over a glazed lightwell. The decks are fronted by balustrades of bolt-fixed glass and are set back to maintain neighbours' rights to light, giving a swept-back profile of facetted glass a little like **City Hall**. Five teaching floors are raised above an undercroft that accommodates an assembly hall, chapel and sheltered play area. The semi-open basement and the 16m (52ft) span of the lightwell is achieved by means of an exposed bow-arched truss from which cables fan out to pick up the floor loads. A tensile canopy caps the school, sheltering a rooftop technology garden. The design informed BDP's Beehive School, one of several 'exemplar designs' commissioned in advance of the government's Building Schools for the Future programme.[59]

LONDON CITY HALL

110 The Queen's Walk, Southwark, London
2000–2; Foster + Partners
Consultants for structural engineering, services, façade, fire and acoustics: Ove Arup & Partners

On the abolition of the Greater London Council in 1986, the capital was left without a strategic regional authority. After a London-wide referendum, the Greater London Authority (GLA) was established in May 1998, having responsibilities for transport, policing, fire and rescue, development and strategic planning. County Hall had been sold off, so the Labour government procured a new headquarters on a 25-year lease. A deal was struck with the developers of More London, a site being masterplanned by Foster + Partners. They gained an anchor tenant; in return City Hall took a prominent site on a (privately owned) riverside plaza adjacent to London Bridge. Under partner in charge Ken Shuttleworth, the scheme underwent several iterations, dubbed 'box', 'mask' and 'flask', before crystallizing in 1999.

The building resembles a glass egg, tilted north to limit solar gain and stepped to the south to provide self-shading. At its heart was an assembly chamber in the form of a stepped cone, surrounded by a spiral ramp derived from Foster's renovation of the Reichstag for the German parliament. The fully glazed envelope is mitigated by low-energy features such as mixed-mode ventilation, borehole water cooling and grey water recycling, achieving a BREEAM 'Excellent' rating in 2002. Critics hailed City Hall as a symbol of democratic values (transparency, accessibility, sustainability), yet the GLA was simply a tenant. After the body relocated in 2020 (to Wilkinson Eyre's The Crystal at the Royal Docks) City Hall, assembly chamber and all, was stripped out – an unintentional metaphor for devolved government in the neoliberal age. A retrofit is planned.

SENEDD CYMRU

Cardiff Bay, Cardiff
2001–5; Richard Rogers Partnership
Structural engineer: Arup

Located in the former Bute West Dock, the home of the Welsh Assembly government looks out to Cardiff Bay and beyond. Despite a fraught procurement process and post-9/11 security measures, the clarity, openness and legibility of Ivan Harbour's 1998 competition entry just about prevails. Stressing universal values over the particularities of place, he sought a 'natural gathering place' that would welcome a sceptical electorate.[60]

The meeting of land, sky and sea is answered by two dynamic planes: an inhabited slate plinth that steps up from the waterfront and a big, oversailing roof, cantilevered from 12 slender columns. A cowled funnel over the debating chamber breaks through the roofline, introducing natural ventilation and reflected daylight. Like the courtrooms of RRP'S Strasbourg Court of Human Rights and Bordeaux Law Courts, Y Siambr is a non-confrontational circle in response to Jim Callaghan's plea (as chair of the selection panel) for a building that would 'not be overly adversarial in shape or argument'.[61] Housing 60 members, it is an intimate but theatrical space ringed by a public gallery. Above is what Harbour terms a living room for the electorate, a great public interior dominated by the surge of the funnel into the undulating roof, the whole clad in cedar strips. Sustainable features including rainwater harvesting, ground source heat exchangers and a biomass boiler contributed to an 'Excellent' BREEAM rating. The slate floor extends out to form the external landscaping, which features two public art installations, *Assembly Field* by Danny Lane and Richard Harris's *Meeting Place*.

Further reading:

Trevor Fishlock, 2011, *Senedd*. Cardiff: Graffeg Books.

HOBGOBLIN
J.P.Morgan

HOBGOBLIN
LORDS.ORG/EVENTS
ENGAGE
EXPERIENCE THE
AT THE
HOME OF CRICKET
CGI
J.P.Morgan

3 | sport

IBM SPORTS HALL

Hursley Park, Winchester, Hampshire
1981–2; Nicholas Grimshaw & Partners
Structural engineer: F J Samuely & Partners

IBM UK purchased Hursley Park in 1958, adding laboratories and offices in stages (the earliest employing the American industrial designer Eliot Noyes as design consultant). Like IBM **Cosham**, the brief was modest: a low-budget hall for three badminton courts, tucked away behind a timber-framed clubhouse. After initial explorations of fabric structures were turned down by the client, the practice developed its first exoskeleton.

Five portal frames of trussed tubular steel permit a largely unencumbered interior. John Winter, Grimshaw's former tutor at the Architectural Association, praised the clarity and elegance of the black frame, which 'has something of the impact of those wire bases to Eames chairs which arrived on the scene in the mid-'50s', although the frames seem overelaborate for their 18m (60ft) span.[62] The building was clad in horizontal sandwich panels, bolted at each corner to cleats on the frame, joined with a neoprene gasket and caulked with mastic. For the curved corners the architects chose translucent GRP panels, their azure hue recalling IBM'S nickname 'Big Blue'. During the day they admit natural light; at night, they provide slivers of light to those approaching.

One of Grimshaw's first commissions after the split with Terry Farrell, the Hursley pavilion bears family resemblances to Farrell/Grimshaw's industrial units at **Winwick Quay**, Warrington and Gillingham Industrial Park. In 1982 the practice designed a construction system for the Sports Council: the Standardised Approach to Sports Halls (SASH).

Pages 86–7: Mound Stand, Lord's Cricket Ground, London, 1985–7, Michael Hopkins & Partners.

OXFORD ICE RINK

Oxpens Road, Oxford
1983–4; Nicholas Grimshaw & Partners
Engineers: Ove Arup & Partners

Oxford's rink came after a campaign spearheaded by Mary Meagher, who founded the Oxford Ice Skating Club in 1980. Their cause was taken up by Councillor Barbara Gatehouse of Oxford City Council's recreation and amenities committee, which earmarked funding and a riverside site. In 1983 Grimshaw was briefed on a ten-month programme and a requirement for the venue to be as self-financing as possible. It was agreed that the building needed to advertise itself and avoid a warehouse-like appearance.

Grimshaw's team settled on a long-span structure in which an overhanging spine beam is rigged by stainless-steel tension rods strung from twin masts, 33m (110ft) in height. This provided a column-free interior and minimized the need for piling amid poor ground conditions. Most importantly, the structure conferred a suitably striking image; that of a two-masted schooner moored on the Thames (the building was known locally as 'Barbara's battleship'). Masted structures had been applied to ice rinks in Germany (Eisstadion Braunlage, 1972) and Switzerland (Eissporthalle Obersee Arosa, 1978), while Grimshaw's deployed a single mast for their Heatwaves leisure centre at Stockbridge Village, Merseyside (1986–8, demolished 2010). The cladding required good thermal and acoustic insulation properties; the solution took the form of GRC sandwich panels originally developed for cold stores. A triple-glazed north wall provides natural light without compromising internal temperatures. The rink has remained popular with Oxford's skaters and its resident ice hockey team, the Oxford City Stars. But it faces an uncertain future as the Oxpens area undergoes redevelopment.

OXFORD
ICE RINK

LINK CENTRE

Whitehill Way, Westlea, Swindon, Wiltshire
1983–5; Nigel Honer, Thamesdown Borough Council
Structural engineer: Anthony Hunt Associates

Swindon's Link Centre brings together the community's sports and leisure facilities under a big sheltering roof. The facility provides a full-size ice rink, swimming pool, health and fitness suite, sports hall, snooker room, district library, arts centre and theatre, community suite and youth-club room. Swindon was one of the UK'S fastest-growing towns after 1945, with a large area of expansion to the west.

The Link forms the most identifiable part of a district centre that includes a shopping parade, supermarket and health centre. Community, arts and sports facilities were co-located in many of the new and expanded towns, an idea that can be traced back to Henry Morris's Cambridgeshire village colleges. The facility was financed by a land deal that provided a fixed sum, so a fast-track programme was chosen to minimize its devaluation by inflation. A highly prefabricated building suited the intensive schedule: its tubular steel roof frame was erected while the interior layout was still being finalized. The grand structure unifies a jigsaw of small and large spaces while providing a flexible and largely column-free interior. The Link has a square plan with a corner snipped off to denote the entrance.

Like the **Renault** building down the road, it is a masted structure. From an array of perimeter columns and outriggers are suspended steel tension hangers. These support the roof deck, a two-way grid of lattice girders. Services are exposed and integrated into the frame, with plant housed in a rooftop tower.

J.P.Morgan
J.P.Morgan
J.P. Morgan Media Centre
Health and
Life insurance

LORD'S CRICKET GROUND

St John's Wood Road, London
Mound Stand: 1985–7; Michael Hopkins & Partners
Structural and services engineer: Ove Arup & Partners
Media Centre: 1997–9; Future Systems
Structural engineer: Ove Arup & Partners

Lord's marks the moment of Hopkins' embrace by the Establishment; after this came Glyndebourne Opera House, Emmanuel College Cambridge, Buckingham Palace and Sherborne School. To celebrate their bicentenary, the Marylebone Cricket Club decided to rebuild Thomas Verity's 1898–9 Mound Stand. Its replacement (pages 86–87) layers a summer tent over steel over brick in a built diagram of social stratification. Verity's rear arcade was extended and the terrace refurbished. Above this is the world of corporate hospitality: private boxes and dining rooms, debenture seats and restaurant. These are housed in a tiered steel superstructure, cantilevered from a girder supported on six columns. The upper promenade is crowned by a translucent fabric roof that, like a village marquee, provides just enough shelter and no more.

Lord's seems to have gained a taste for adventure, commissioning Nicholas Grimshaw & Partners to replace Herbert Baker's Grand Stand in 1996–8. At the Nursery End is the Media Centre, a futuristic press box inspired by high-tech athletics garments and the plastic styling of a digital camera. The Media Centre is a streamlined pod raised 14m (45ft) on a pair of elephantine service tower legs. The semi-monocoque capsule was assembled in Falmouth's Pendennis Shipyard from bolted and welded aluminium sheet, ribbed and sparred for rigidity, then sanded and sprayed to remove any artefact of its construction. In 2015–17 it was refurbished by David Miller Architects (Miller being the original project architect). This was succeeded by Wilkinson Eyre's rebuilding of the Compton and Edrich Stands (2019–21), to be followed by the Tavern and Allen Stands.

Further reading:

David Jenkins, 1991, *Mound Stand, Lord's Cricket Ground*. London: Architecture Design and Technology Press.
Jan Kaplický, 2001, *Future Systems: Unique Building: Lord's Media Centre*. Chichester: Wiley-Academy.

KIRKLEES STADIUM

Stadium Way, Huddersfield, West Yorkshire
1993–4, 1996–8; Lobb Partnership
Structural engineer: YRM Anthony Hunt Associates

A crowd crush at the FA Cup semi-final at Sheffield's Hillsborough Stadium on 15 April 1989 resulted in the loss of 97 lives. Lord Justice Taylor's inquiry into the disaster recommended all-seater stadiums, precipitating a new era in British football. Huddersfield's Leeds Road ground was one of scores that failed to meet the Taylor recommendations. Looking to relocate, the club formed a consortium with Huddersfield Rugby League Club and Kirklees Metropolitan Council, who provided a site between the River Colne and the wooded Kilner Bank. Funding of £15m was assembled from grants and sponsorship but it was clear that the ongoing viability of the venue would depend upon corporate hospitality and commercial developments.

In 1990 a prototype 'Stadium for the Nineties' came to the club's attention. It was devised by Geraint John of the Sports Council, Rod Sheard of the Lobb Partnership and Stephen Morley, a structural engineer who ran Tony Hunt's Sheffield office. They drew an oval around a pitch and filled the four segments with stands that could be put up in phases. Lobb and Hunt were invited to develop the design in a limited competition. Their layout is based on a maximum viewing distance of 150m (500ft) from the four corners of the pitch. The unifying gesture comes from the curved 'banana trusses' of white tubular steel that meet at raking concrete finger supports. Developed from Hunt's prismatic trusses at **Waterloo International**, the long arches span 135m (440ft) but weigh only 78 tonnes. Their progeny can be seen at many later stadia, including Morley's roof for the Sydney Olympic Stadium (1996–9, HOK Sport).

VENARI

LONDON REGATTA CENTRE

1012 Dockside Road, Royal Albert Dock, Newham, London
1997–9; Ian Ritchie Architects
Structural engineer: Ove Arup & Partners

The Royal Albert Dock Trust was formed in 1992 by the Olympic rower Fred Smallbone to support community water sports at the former docks, then undergoing regeneration. The London Docklands Development Corporation granted a 200-year lease for a triangular site at the north-west corner, hemmed in by the Connaught Bridge and the elevated viaduct of the Docklands Light Railway. When Smallbone approached the Richard Rogers Partnership, Mike Davies passed him on to Ian Ritchie – both architects sat on the LDDC'S urban design advisory board. Ritchie in turn brought on board an Arup team that included an experienced rower, James McLean, as project engineer.

A two-storey clubhouse and boathouse overlook a 2,000m (6,500ft), seven-lane rowing course. Both structures hug 3.5m-high (12ft) freestanding retaining walls in the form of vertical 'pillow' gabions: wire mesh cages filled with granite rubble. Ritchie introduced structural gabion walls at the cultural greenhouse (1992–4) at the Jardins de l'Imaginaire in the Dordogne. At Albert Dock they provide stability, protection and a tough presence in the open landscape. The clubhouse is shaped like a Stanley knife, with a 90m (300ft) 'blade wall' of fair-face concrete set parallel with the dock wall. Inside is a powered rowing tank, developed by Arup to simulate open-water conditions. A cantilevered upper terrace provides views of races and planes taking off from London City Airport opposite. The clubhouse was extended around 2016 without Ritchie's involvement. The structural columns of the boathouse support 'eights', 'quads' and 'pairs' boats as well as a repair workshop. Four galvanized sheet steel catenary roofs enclose the space between parallel gabion walls.

QUEEN ELIZABETH OLYMPIC PARK

Newham, London
London Stadium: 2008–11; Populous (formerly HOK Sport)
Structural engineer: Buro Happold
Lee Valley Velodrome: 2009–11; Hopkins Architects
Structural engineer: Expedition Engineering

Each Olympic games embodies a different tension between a festival of sports and the provision of permanent infrastructure. London 2012 contrived a balance between 'legacy' – including a new park – and temporary facilities that earned it the nickname of 'the flat-pack Olympics'.[63] Flexibility served the Olympic Stadium well. The original brief called for a 25,000-seat permanent athletics venue, a concrete bowl dug into the ground and surrounded by waterways. To this was added a demountable upper stand, providing an additional capacity of 55,000 for the duration of the games. A superstructure of bolted steel tubes (repurposing surplus gas pipelines) supported a cable-net roof. After 2012, plans changed. The upper stands stayed; they were joined by a new roof, facilities and retractable seating, creating a multi-purpose venue that could be reconfigured for football (West Ham United being the main tenants), athletics, rugby, baseball and concerts.

The Velodrome has changed little since the games, which Hopkins Architects regarded as a housewarming party. Senior partner Mike Taylor wanted to shrink-wrap the stadium around the track, like a Lycra-clad cyclist. This generated a taut, dynamic exterior, clad in red cedar and topped by a cable-net roof curved in two directions like a Pringles crisp. The upper tier is raised 3m (10ft) above a continuous glazed strip, connecting inside and out. Both structures are lightweight and high performance, taking their cues from the athletes they serve. Zaha Hadid Architects' London Aquatics Centre (2008–11) embodies a very different relationship between form and technique – its roof structure uses 30 times as much steelwork as the Velodrome's to cover a similar area.

Further reading:

Dave Hill, 2022, *Olympic Park: When Britain Built Something Big*. London: On London Books.

4 | culture and
entertainmen

SNOWDON AVIARY

London Zoo, Regent's Park, London
1962–5; Lord Snowdon, Cedric Price
Structural engineer: Frank Newby of Felix Samuely and Partners
Listed at grade II* in 1998

What better place to see zoomorphic architecture than London Zoo? Just as the textured masses of Casson and Conder's Elephant and Rhino Pavilion evoke its ponderous inhabitants, so its exact contemporary, the Snowdon Aviary, appears to hover in mid-air. Britain's first walk-through aviary is a cat's cradle of tubular aluminium, welded mesh and high-strength cables. Between two pairs of tetrahedra are interposed a pair of V-shaped legs, anchored to the ground with rocker bearings. This armature is kept in tension with cables, and around it is thrown a net of black anodized aluminium. Through the cage snakes a cantilevered walkway intended to provide views of nesting birds on the terraced site. The structure was designed on the 'tensegrity' or tensional integrity principle developed by R Buckminster Fuller, with whom both Price and Newby were in touch. Isolated elements are held within a network of pre-stressed tension members.

The aviary is redolent of the pavilions of world fairs and expos, with their peculiar combination of novelty, levity and fantasy. The obvious precursor is the Skylon at the 1951 Festival of Britain, a tensegrity structure on which Newby cut his teeth. As structural engineer, Newby seems to have acted as the lynchpin of the design team, devising in large part the structural concept. The framework employed aluminium castings and forged stainless-steel components developed for use in industrial plant and the aircraft industry. Fun, freedom, simplicity of concept and complexity of elaboration – these qualities were reprised at the Fun Palace (another Price/Newby collaboration) and the Centre Pompidou. In 2020–2 the Aviary was conserved and repurposed as 'Monkey Valley', housing ten eastern black and white colobus monkeys.

Pages 104–5: Great Glasshouse, National Botanic Garden of Wales, Carmarthenshire, 1997–9, Foster Associates.

FRONT RANGE GLASSHOUSES

Royal Botanic Garden Edinburgh, Inverleith, Edinburgh
1965–8; Ministry of Public Buildings and Works, Scotland
(G A H Pearce, architect in charge)
Structural engineer: MPBW Scotland (L R Creasy, F W Walley, H E Mills)
Listed at grade A in 2003

Due to their unique technical challenges, glasshouses have the potential to be hotbeds for constructional innovation. The masted glasshouses at Inverleith are high-tech pioneers – yet their structural principles were anticipated in 1853 by Charles McIntosh, head gardener at Dalkeith Palace.[64] They were the brainchild of E E (Eddie) Kemp, curator of the Royal Botanic Garden Edinburgh (RBGE), who in 1962 secured funding to replace six Edwardian display houses. He wanted unobstructed spans for an expansive, naturalistic landscaping scheme. The 128m-long (420ft) exhibition house was divided into five climatic zones, hosting cacti and succulents, warm temperate aquatic plants, temperate plants, tropical aquatic plants and temperate ferns. A shorter orchid house joins at right angles. Informal routes wound around the planting; a raised walkway allowed visitors to inspect the crowns of specimens, while the underside of the giant waterlily could be viewed from an underground chamber.

The MPBW team started with a three-pinned portal frame with sloping sides, but it could not be made to withstand the various structural forces, including wind and snow loads. Tetrahedrons were added, not suspended in mid-air as at the **Snowdon Aviary**, but as a row of space frames supporting the glazed envelope. They are stiffened with struts, while stainless-steel cables suspended from the ends relieve the rafters of some of the structural loads. The tensile framework is made up of welded hollow steel sections – tubes for the external structure and rectangular sections for the roof – on a reinforced concrete base. Pearce adapted the structure for his display greenhouse at Chatsworth (1970). The Inverleith glasshouses are undergoing restoration as part of the RBGE's Edinburgh Biomes project.

Further reading:

David Rae, 2011, *The Living Collection*. Edinburgh: Royal Botanic Garden Edinburgh.

ROYAL EXCHANGE THEATRE

St Ann's Square, Manchester
1975–6; Levitt Bernstein Associates; theatre design consultant: Richard Negri
Structural and fire safety engineer: Ove Arup & Partners
Former Royal Exchange listed grade II in 1974

The 1970s was a heady moment in theatre design; alongside behemoths such as the National Theatre thrived a counterculture that valued independence, impermanence and improvisation. In 1971 the theatre directors Casper Wrede, Michael Elliott and Braham Murray took a 21-year lease on the Royal Exchange, where trading had ceased in December 1968. With the production designer Richard Negri and project architect Malcolm Brown they devised a 'theatre module' that would touch down in the cavernous space of the cotton hall. Like the Round House in Camden and Liverpool's Everyman, it adapted a non-purpose-built (and historic) 'found space'.

What was unexpected was the radicalism of their architectural intervention, an openwork pod of tubular steel poised under the central glazed dome and partly suspended from its scagliola piers. It was envisaged as 'a Chinese lantern lighting up an old dark place', although the Arts Council's officer sniffed that it 'resembled one of the American moon landing craft and is probably just about as technically sophisticated and nearly as expensive'.[65] Inside is a galleried, in-the-round playhouse with entrances shared by audience, actors and crew. The auditorium is seven-sided so that no member of the audience sits directly opposite someone else. There is raked seating for 400 in the stalls with two galleries each seating 150. Louvres within the roof can be adjusted to calibrate the auditorium's reverberation time, while the translucent cladding invites imaginative lighting effects. After the venue was damaged in the 1996 Manchester bombing, Levitt Bernstein were brought back to introduce improvements.

Further reading:

Alistair Fair, 2018, *Modern Playhouses: An Architectural History of Britain's New Theatres, 1945–1985*. Oxford: Oxford University Press.

SAINSBURY CENTRE FOR VISUAL ARTS

Norfolk Road, University of East Anglia, Norwich, Norfolk
1977–8, 1989–90, 2004–5; Foster Associates
Structural engineer: Anthony Hunt Associates
Listed at grade II* in 2012

After Sir Robert and Lady Lisa Sainsbury donated their extraordinary collection of modern and ethnographic art to the University of East Anglia, they added an endowment for an art museum that would keep it intact and open it up for people to enjoy. Still later it was decided to integrate the teaching and library facilities of the School of Fine Art, these elements being funded by the University Grants Committee. Faced with two very different clients and an overloaded brief, Foster played a series of decisive moves. It would have been easy to decouple the sets of users into linked pavilions, but his instinct was to pack everything into a single box and deal with the consequences. From the start, it would seem, this took the form of an open-ended tube, like an enlargement of the **Modern Art Glass** building. This was sliced into a series of zones for the storage and display of art, visitor facilities and academic functions. Foster's preferred site lay on the western fringe of the campus, keeping his distance from Denys Lasdun's megastructures to frame views of Brenda Colvin's landscaped parkland.

At the eleventh hour, faced with the realization that 'there was too much stuff compromising the space', he overhauled the plan, introducing a pair of mezzanines and a basement spine.[66] But the real breakthrough was the decision to transform the cladding into an inner and an outer skin, prised far enough apart to accommodate the services and cubicles that threatened his open plan. Tony Hunt redesigned the structure, devising a series of space trusses, forming big triangles in section. These are shown off at the glazed ends of the tube and glimpsed from inside through a screen of white louvres.

Perhaps the ultimate refinement of the tin shed in a green field, the Sainsbury Centre announces the quiet sophistication and cultural ambitions that would become Foster hallmarks. It is true that his layout privileges some spaces and users over others – one is reminded of Paul Rudolph's remark that 'Mies makes wonderful buildings only because he ignores many aspects of a building' – but the exhibition spaces, dappled with sunlight filtered through a curtain of

white blades, are on a par with Renzo Piano's Menil Collection and Alvar Aalto's Kunsten Museum of Modern Art – the last jointly visited by architect and client.[67] For Peter Cook, the building lacked the tension and single-mindedness of **Willis Faber & Dumas**; in its eagerness to reveal structure and services, 'the potential *absoluteness* of this building is eschewed in favour of the old game of "event"'.[68] Cook surmised that Sainsbury would become more of a monument than a well-serviced shed, and he was proved right.

When Foster was called back to add storage, curatorial and conservation facilities, he chose not to further extrude the tube but to fan out its basement, which emerges downslope with a slash. In section it recalls one of Foster's earliest projects, The Cockpit (1965), a retreat built at Pill Creek for Marcus Brumwell. A new public route to the Crescent Wing was added in 2004–6, when the Centre underwent refurbishment.

Further reading:

Witold Rybczynski, 2011, *The Biography of a Building: How Robert Sainsbury and Norman Foster Built a Great Museum*. London: Thames & Hudson.

WILTSHIRE RADIO

Lime Kiln Studios, Wootton Bassett, Wiltshire
1982–3; Nicholas Grimshaw & Partners
Structural engineer: F J Samuely & Partners

While Terry Farrell's TV-am and Limehouse Studios were conversions of industrial buildings, his erstwhile partner designed a tiny radio station in a walled garden. The first licences for commercial radio were issued in the 1970s, leading to a boom in local independent stations. Wiltshire Radio went on air in October 1982 and their studio was completed the following year. With an area of 170m² (1,800ft²) and a contract value of £198,000, it is one of Grimshaw's smallest projects (the equivalent figures for the contemporaneous **Herman Miller Distribution Centre** in Chippenham are 6,970m² (75,000ft²) and £1.68m). Early elevations presented the building as a transmitter radio, complete with telescopic aerial.

The 'space ship' (as Grimshaw described it) accommodated a lightly framed reception area and a heavily insulated studio, while the record library and offices remained in the late 17th-century Lime Kiln House.[69] The location was a noisy one, on a main road and under an RAF flight path. To improve the acoustic insulation, the studios (two live studios, one talks studio, a news booth and a production suite) were planned as six cells of cavity blockwork lined with profiled foam and clad with an 'acoustic jacket' of GRP panels. A glazed spine link is cranked out to form an entrance canopy. On installation the panels, ribbed with curved corners, were sealed with neoprene gaskets then filled with sand to improve their acoustic properties. The facility later became a business centre; at the time of writing there are proposals to convert the former studios to a nursery.

PRIF FYNEDFA
MAIN ENTRANCE

TECHNIQUEST

Stuart Street, Cardiff
1993–5; Ahrends Burton and Koralek
Civil, structural and services engineers: Buro Happold

Launched in temporary premises in 1986, Techniquest was the UK'S first science discovery centre. Millions of young people have since engaged with science and technology through hands-on exhibits. It was the brainchild of Professor John Beetlestone, a 'retailer of science' who witnessed interactive learning at the Evoluon in Holland, the Ontario Science Centre and the Exploratorium in San Francisco. In 1991 Cardiff Bay Development Corporation selected Techniquest as a lead project in the regeneration of the former Cardiff Docks. When the late 19th-century repair workshop of the Mount Stuart Dry Dock was put forward as a temporary exhibition space, Paul Koralek suggested that its iron frame could make the bones of a permanent venue. As at ABK'S Dover heritage centre, a retained element becomes the foil to a series of playful interventions.

The barrel-vaulted shed became an open-plan exhibition hall, animated by a gallery, information pod and restored gantry crane. It is a tight, tall space, with joyful shrieks echoing off its hard surfaces. The initial group of 160 exhibits, in glossy primary colours like a juvenile Bauhaus, were arranged with 'carefully thought-out randomness'.[70] Structural glass walls, fixed by an external scaffold of tubes and tension rods, make garrulous companions to their Victorian ancestors. Ancillary elements spill outside: a curved workshop studio, an apsidal demonstration theatre (inspired by London's Royal Institution Theatre) and 'Starlab', an aluminium-clad geodesic dome. In 2019–20 the workshop block was replaced by a 'Science Capital' extension by HLM Architects, which incorporated a new entrance.

AMERICAN AIR MUSEUM, DUXFORD

Imperial War Museum, Duxford Airfield, Cambridgeshire
1995–7; Foster Associates
Structural engineer: Ove Arup & Partners
Listed at grade II* in 2020

Duxford aerodrome was established as a training depot station in 1917–18, when the first arrivals included members of the American Expeditionary Force sent to fight in the First World War. Duxford became an important fighter station in the Battle of Britain and was later occupied by the US Eighth Air Force. After the closure of the base in 1961 the Imperial War Museum took it over for the storage, restoration and display of a growing collection of American combat aircraft. Foster's hangar is no tin shed but a twin-skin concrete shell. Geometrically, it is a section of a torus (inner tube), segmentally curved in two dimensions like **Canary Wharf Station**. Its two constant radii meant that 924 prefabricated roof panels could be assembled from six sets of components. Early on, Arup's weighed up steel and concrete roofs, concluding that while it was slightly more expensive in construction, the latter would be more thermally stable and less prone to condensation, requiring only minimal dehumidification to attain museum standards.

The exhibition hall emerges from a grass bank, overlooking the runway through a glass wall. The space, with its 90m (300ft) span, was designed around the largest exhibit, a Boeing B-52 Stratofortress, with other planes hung from the roof like Airfix models. Visitors enter at the back via a splayed tunnel resembling the entrance passage to a prehistoric tomb. They emerge to be confronted nose-to-nose with the B-52. Twin ramps curve around the shell, illuminated by a glazed slit like the Crescent Wing at the **Sainsbury Centre**. In March 2016 the American Air Museum was relaunched with a new exhibition design by Redman Design.

BFI LONDON IMAX CINEMA

1 Charlie Chaplin Walk, South Bank, London
1997–9; Avery Associates
Structural engineer: Anthony Hunt Associates

Bryan Avery (1944–2017) was a visionary thinker whose fascination with advanced technology was allied to an abiding interest in the culture and experience of urban life. This urban cinema was a sequel to his Museum of the Moving Image (1985–7), slotted under the Waterloo Bridge approach. The British Film Institute's new site was scarcely more promising: a sunken island at the centre of the Bullring roundabout, accessed from a labyrinth of underpasses. Early designs were based on a metal drum covered with greenery (hinting at Avery's concept of a 'wilderness city'). As built, the planting was pruned back to a green plinth, revealing an 'image wall' encased in glass. Avery compared the rotunda form, with its high-level slots, to a zoetrope or a slide carousel.

IMAX, its name derived from 'maximum image', is a large-screen motion picture format developed in Montreal in the late 1960s. The cinema is entered at underpass level, with cranked stairs ascending to a foyer. Above is the auditorium, with steeply raked seating focused on the UK'S biggest screen (20.4 × 27.6m; 67 × 90ft). The outer skin of structural glazing is suspended from 'gallows brackets' cantilevered from the main columns. Onto the 'image wall' was pasted a wraparound enlargement of a Howard Hodgkin painting (replaced in 2006 by advertising). The Waterloo & City Underground line runs 4.5m (15ft) below the building, requiring a complex foundation design and spring bearings to prevent vibrations being transmitted to the auditorium above. In 2021 it was reported that freeholders Lambeth Council and Transport for London had commissioned a feasibility study for the redevelopment of the IMAX with an office tower.[71]

Further reading:

Bryan Avery, 2011, *Fragments of Wilderness City: The Work of Bryan Avery*. London: Black Dog Publishing.

BFI
IMAX
save with us
e·on
next
IMAX
OCEAN

MILLENNIUM DOME

The O2 Arena, Peninsula Square, North Greenwich, London
1997–9; Richard Rogers Partnership
Engineer: Buro Happold

Long before the show came to town, the big top went up. An initial design study by Gary Withers of **Imagination** fitted a circular layout within the big meander of the Thames at Blackwell Point, then being masterplanned by RRP for English Partnerships. In Mike Davies's sketch of May 1996 this became the familiar image of a dome suspended from a ring of masts. An all-encompassing tent offered several advantages over a dozen expo pavilions – it was inexpensive, offered a singular image and sheltered visitors from the bitterly exposed site. Most importantly, the high-tech tactic of decoupling packaging from content bought time for those devising the 'Millennium Experience'. In structural terms the Dome is not a dome but a cable-net structure – a fabric membrane pulled taut by tension cables. They are lashed down to a scalloped edge, allowing multiple points of entry.

This is a festive marquee, combining two structures that had fascinated a young Davies at the 1951 Festival of Britain: Ralph Tubbs' Dome of Discovery and the Skylon, designed by Powell and Moya with the engineer Felix Samuely. Louvres within the Dome's central cap vent the 2 million cubic metres (70 million cubic feet) of enclosed air, driven by exhaust fans within the 12 service cylinders circling the perimeter. A cut-out in the Teflon-coated roof uncovers the southern ventilation tower of the Blackwall Tunnel. In 2005–6, the 20,000-seat O2 Arena was erected inside. Served by **North Greenwich Station**, it has, unlike the Millennium Experience, been a conspicuous commercial success – millennials are more likely to know the Dome as the O2.

Further reading:

Elizabeth Wilhide, 1999, *The Millennium Dome*. London: HarperCollins.

GREAT GLASSHOUSE, NATIONAL BOTANIC GARDEN OF WALES

Middleton Hall, Llanarthne, Carmarthenshire
1997–9; Foster Associates
Structural engineer: Anthony Hunt Associates
Landscape architect: Gustafson Porter

The UK'S youngest national botanic garden is situated in the rolling hills of rural Carmarthenshire. After Middleton Hall was devasted by fire in 1931 its 18th-century parkland gradually ran to seed. From 1989 the Welsh Historic Gardens Trust campaigned to restore the 568-acre estate as the home of a National Botanic Garden of Wales. A landscape plan by Hal Moggridge connects aspects of the historic landscape, including a chain of five lakes, with woodlands and wildflower meadows.

The largest single-span glasshouse in the world, Foster's dome is the focal point. It blisters out of its surroundings, an unearthly presence in the lush landscape. On plan the Great Glasshouse is an ellipse, echoing the round platform on which Middleton Hall stood. Like **Duxford**, this 4,500m^2 (48,500ft^2) glass roof is a sliver of a gigantic torus (or doughnut), tilted at seven degrees towards the south-west for sun and views. The roof is spanned by 24 tubular steel arches to which are welded T-sections that support the aluminium glazing system. The arches connect, via ball-and-socket joints, to a concrete perimeter ring beam that caps an embanked substructure. Three tunnel entrances lead to a sunken interior. Kathryn Gustafson's planting scheme, now mature, focuses on flora from Mediterranean climates around the world. Ventilation comes from 147 computer-controlled opening panes, while a perimeter heating system is fed by nearby biomass boilers. Runoff from the dome is collected into underground tanks for irrigation.

Further reading:

Andrew Sclater, 2000, *The National Botanic Garden of Wales*. London: HarperCollins.

EDEN PROJECT

Bodelva, Par, Cornwall
1998–2001; Nicholas Grimshaw & Partners
Engineer: Anthony Hunt Associates

In its fusion of technology and ecology, the Eden Project evokes the thinking of R Buckminster Fuller, one of Nicholas Grimshaw's early heroes. Two clutches of linked geodesic domes rest on an undulating landscape like clusters of soap bubbles on washing-up. Inside are two biomes, one simulating a Mediterranean climate, the other a tropical rainforest. Eden was co-founded by the Dutch-born businessman Tim Smit who, inspired by Conan Doyle, hankered after a 'lost world' of biodiversity. He started by restoring a historic garden at Mevagissey nearby as the Lost Gardens of Heligan.

Only after Grimshaw came on board did Smit find his lost world, a worked-out china clay quarry near St Austell. Sunk 15m (50ft) below sea level, it resembled a lunar landscape. Mining continued until funding was secured, so the terrain was constantly in flux. An initial scheme based on the **Waterloo Terminal's** variable arches was dropped in favour of a more organic and lightweight solution. The structure, developed with Tony Hunt, is based on a double-layered Mero space frame. Steel tubes are connected by a universal ball joint of cast steel. An outer layer of hexagons is fixed with diagonal struts to an inner frame of triangles and hexagons. Each dome is of varying diameter but all are set out from the same notional centre line, while spheres of different sizes meet at prismatic arch trusses. The framework is clad in translucent cushions of inflated ETFE foil up to 2m (6.5ft) deep, creating a featherweight structure. Other Edens are projected for Morecambe, Dundee and further afield.

Further reading:

Hugh Pearman and Andrew Whalley, 2003, *The Architecture of Eden*. London: Transworld.

NATIONAL SPACE CENTRE

Exploration Drive, Leicester
1999–2001; Nicholas Grimshaw & Partners
Structural and services engineer: Ove Arup & Partners

Leicester University has been a centre of space research since 1960, when Professor Ken Pounds founded the Space Research Group. The National Space Centre lies to the north of the city, on the banks of the River Soar. Like Leicester University's Engineering Building, it takes the form of a tower and podium. A two-storey exhibition pavilion is sunk within the footprint of an underground stormwater tank (belonging to the former Abbey Pumping Station). Its openings are concealed behind a perforated metal screen. The white geodesic dome of the Sir Patrick Moore Planetarium breaks though its flat roof, resulting in an 'egg in a box' plan.

The 41m-tall (135ft) rocket tower enables visitors to inspect the different stages of two prize rockets, Blue Streak (the first stage of the Europa satellite launch vehicle) and the American Thor-Able. Two curved legs of tubular steel extend up and over a reinforced concrete core, braced by horizontal ribs. They support a lightweight sheath of inflated bands of translucent ETFE film, earlier deployed at the **Eden Project**. At the back, outsized stainless-steel tubes are proudly displayed, as at Grimshaw's Xerox Research Centre in Welwyn (1987–8).

The result is a showcase that partakes of the language of space technology while stopping short of *Star Trek*. More recently, Space Centre has been joined by Space Park Leicester, a hub for space research, with further 'high-tech business incubator facilities' in the form of the Dock and Dock 2 workspaces. Along with the University's Institute for Space, the R&D cluster has been branded as 'Space City'.

Welcome
Presented
BT

LONDON EYE

Jubilee Gardens, York Road, Lambeth, London
1999; Marks Barfield Architects
Structural and services engineer: Ove Arup & Partners

'Why should people always have to look up?' asked Cedric Price in 1984, imagining an observation wheel at Jubilee Gardens.[72] Realized some 15 years later, the London Eye combines pleasure (a ride in the air) with instruction (an accessible overview of the capital and Parliament). The brainchild of David Marks and Julia Barfield with Jane Wernick of Arup's, the project is a reinvention of the 1893 wheel erected in Chicago by G W G Ferris Jr. To permit unobstructed views, the Eye's 32 passenger capsules are mounted outside the rim and the whole structure cantilevered over the Thames.

The Eye works on the structural principle of a spoked bicycle wheel: prestressed steel cables connect a central hub to a tubular space truss rim. Unlike a bike, the hub and spindle assembly is hung from a single A-frame, restrained by stay cables anchored into the ground. The 32 capsules, designed by Nic Bailey, are motorized for stability, enabling passengers to move about freely. One revolution takes about 30 minutes, stately enough to allow passengers to embark and disembark while the wheel turns. Like the Eiffel Tower, the London Eye belongs to the tradition of expo buildings, hovering ambiguously between transience and monumentality. Although it has long joined Big Ben and Tower Bridge on tourist itineraries and memorabilia, only in May 2024 did Lambeth Council approve its permanent retention.[73]

Further reading:

Steve Rose, 2007, *Eye: the Story Behind the London Eye*. London: Black Dog Publishing.

GLASGOW TOWER

Glasgow Science Centre, 50 Pacific Quay, Glasgow
1999–2001; Richard Horden Associates (concept architect),
Building Design Partnership (executive architect)
Engineers: Buro Happold (structural and mechanical),
Peter Heppel Associates (aerodynamics)

In 1992, Richard Horden Associates won an ideas competition for a millennium tower for Glasgow's St Enoch Square. While most towers are engineered for wind loads, Horden had the idea of turning the entire structure into the prevailing wind like a weathervane, reducing aerodynamic drag and the weight of steel required. A cabin at 100m (330ft) would provide dynamic views of the city while a glass-roofed restaurant was incorporated within a circular tower base. The tower's teardrop section acted as an aerofoil, with tapered outrigger wings minimizing turbulence at the trailing edge. Its blade profile was inspired by an upturned helicopter blade and Brâncuși's *Bird in Space* sculptures (1932–40), while Horden compared the bearing rings of the turntable mechanism to the battleship turrets once turned out at the Clydebank shipyards.

The Enoch Square plan came to naught but its promoter, the Glasgow Development Agency, incorporated Horden's Wing Tower into a competition brief for a national science centre. It formed part of Pacific Quay, the regeneration of Cessnock Dock on the Clyde's south bank. BDP won the competition with a Science Centre and IMAX cinema clad in curved titanium sheet. The tower was constructed to a modified design by BDP and Buro Happold and opened in June 2001. Recurring technical faults have resulted in long periods of closure yet it has become a Clydeside landmark and a visitor attraction. At 127m (417ft) in height, the Glasgow Tower is Scotland's tallest, although as an observation tower it has been overtaken by the 162m (530ft) Brighton i360 (2014–6, Marks Barfield) and Portsmouth's 170m (560ft) Spinnaker Tower (2001–5, HGP Architects).

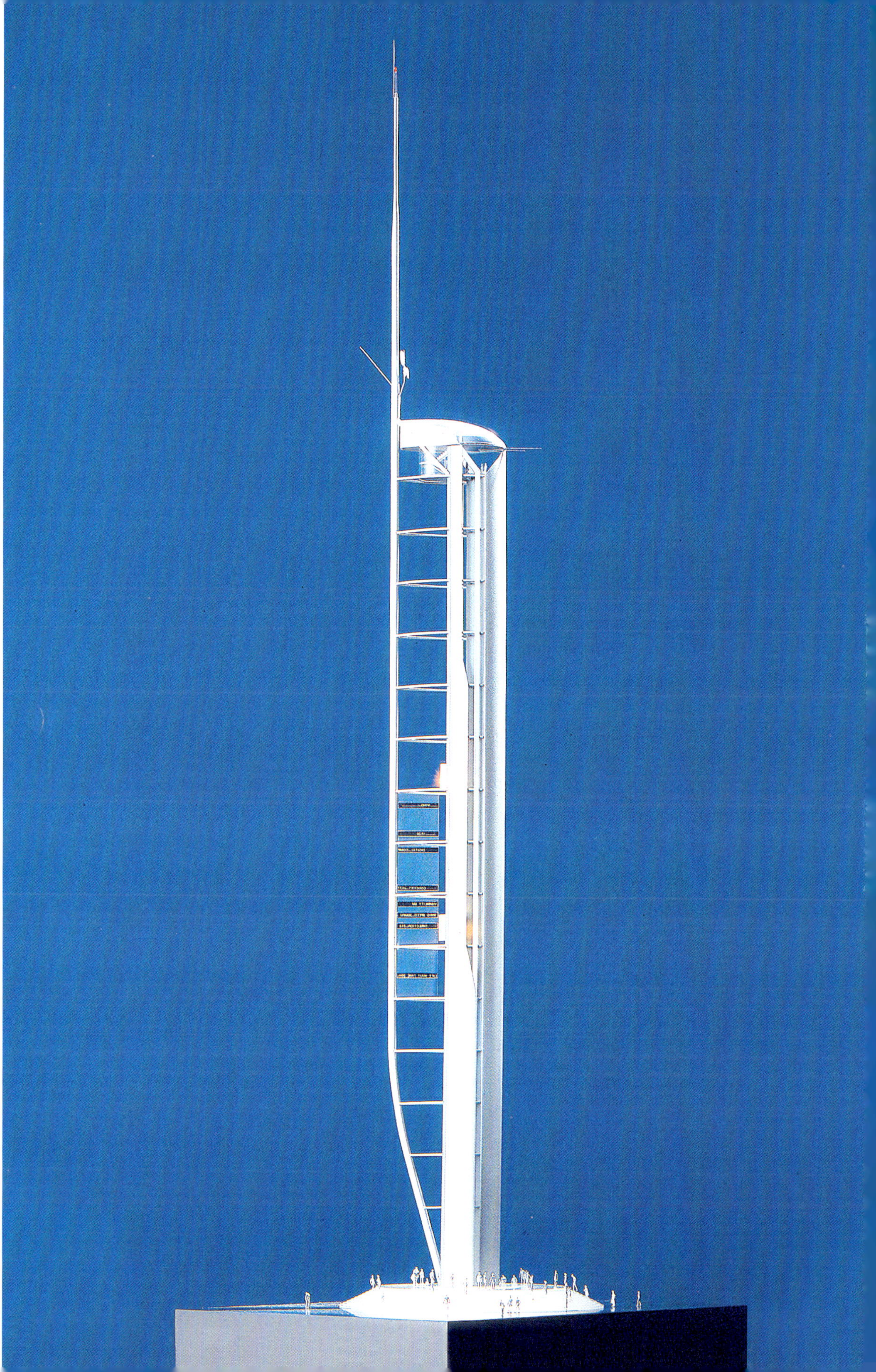

THE GLASSHOUSE INTERNATIONAL CENTRE FOR MUSIC

St Mary's Square, Gateshead Quays, Gateshead, Tyne and Wear
2002–4; Foster + Partners
Structural engineer: Mott MacDonald, Buro Happold

The Glasshouse (formerly Sage Gateshead) and the Baltic Centre for Contemporary Art are the twin lynchpins in the remaking of Tyneside as a cultural centre. The venue is the home of the Royal Northern Sinfonia, the UK'S only full-time chamber orchestra. Foster's first building for the performing arts, the Glasshouse is in some ways an untypical work: a tight-fit layout whose exterior only vaguely gestures at what is going on within. Two performance halls and a smaller rehearsal room are each housed within an acoustically sealed concrete enclosure. The halls sit side by side, linked by a continuous waterside route, so that people are tempted in by the sound of informal musicmaking. Floating over the three music boxes is an undulating, silvery roof, its humped form rhyming with the arches of the Tyne Bridge, the Swing Bridge and the Gateshead Millennium Bridge. The roof terminates in glazed end walls, while scoops at the back create loading bays and service areas.

Each auditorium was allowed to take its own shape under the sheltering roof, aided by Arup's acoustic modelling. The 1,700-seat Hall 1 follows the 'shoebox' layout of the Vienna Musikverein and the Amsterdam Concertgebouw, with tiers of curved galleries overlooked by raked stalls. With a capacity of 450, Hall 2 is a ten-sided galleried space capable of reconfiguration from open stage to in-the-round configurations. The roof structure is made up of four asymmetrical arches whose profile describes a lazy spiral. Against these, curved purlins climb and dip over the halls. They are locked into place by strut and tie members, creating a gridshell structure.

Further reading:

Anthony Sargent and Peter Buchanan, 2010, *The Sage Gateshead*. London: Prestel Publishing.

OVO HYDRO

Exhibition Way, Stobcross Road, Glasgow
2011–13; Foster + Partners
Structural and services engineer: Arup

Glasgow's derelict Queen's Dock was chosen in 1979 as the site of an exhibition and conference centre, now known as the Scottish Event Campus (SEC). Of note are the recent additions to the complex by Foster + Partners: the Clyde Auditorium (1995–7, renamed the SEC Armadillo) and the Scottish Hydro Arena (now the OVO Hydro). The latter provides a 12,500-capacity indoor arena, with the flexibility to meet the sophisticated staging arrangements of stadium tours as well as other events (the venue hosted gymnastics, boxing and netball at the 2014 Commonwealth Games). Sight-line analysis was used to provide the best viewing angles from each seat. This led to a slightly elliptical footprint, with a crescent-shaped seating bowl that rises towards the short axis.

The roof is a shallow dome 123m (400ft) in diameter, tilted to accommodate the inclined bowl. Its diagrid structure is made up of 16 steel trusses connected with cast steel nodes, their lower chords painted light grey to pick out a spiral pattern. Below, radial fins of in-situ concrete support the raked seating and the trusses, which are mounted on steel expansion bearings. The public and service entrances are incorporated into an undulating plinth, like a hovercraft skirt. A Michelin Man façade is made up of bands of inflated ETFE cushions. At night the translucent skin is backlit by LEDS to create a pulsing electronic 'bill-ding board' worthy of Robert Venturi.[74] A more recent source for the plinth and illuminated façade is Herzog & de Meuron's Allianz Arena in Munich (2003–5).

MILES STAIR, SOMERSET HOUSE

Strand, Westminster, London
2013; Eva Jiřičná Architects
Structural engineer: Techniker

More than any other architect, Eva Jiřičná is associated with the sumptuous staircase. One of her earliest flights of fancy was at the Joseph Store at 26 Sloane Street, Westminster (1989). After the store was refitted, this virtuoso assemblage of glass treads and slender steel rods was put into storage, before eventually being shipped to Copenhagen where it was recut for the Birger Christensen store. Less ephemeral is this helical staircase at William Chambers' Somerset House. It is named after Gwyn Miles, outgoing director of the Somerset House Trust. In a previous role at the Victoria & Albert Museum, Miles commissioned EJA to design a new jewellery gallery (2008) in which a bijou staircase rises to an inserted mezzanine.

The Somerset House stair formed the centrepiece of a refurbishment of the west wing to provide offices and conference facilities after HM Revenue & Customs moved out. The existing 1960s stairwell couldn't accommodate a passenger lift, so a new stairwell was inserted into an outer bay by removing four floors. Jiřičná envisaged a 'simple and beautiful sculptural object floating freely in the space'.[75] Treads and landings of Ductal ultra high-performance concrete (UHPC) form a flowing spine as matt and white as bone. A pair of pre-stressed bolts links each step to the one below. The treads are suspended from a steel mesh newel, while a network of tension rods stabilizes the structure. The central post is composed of stainless-steel rods, 3mm (0.1in) in diameter, woven around circular bosses. These are tied together by what EJA partner Duncan Webster describes as 'bicycle chain bracing'. Curved balustrades of low-iron glass provide lateral stiffness without competing with the neoclassical setting.

OATCAKES

5 practices

IBM PILOT HEAD OFFICE

Lynx House, 1 Northern Road, Cosham, Portsmouth, Hampshire
1970–1; Foster Associates
Structural engineer: Anthony Hunt Associates
Listed at grade II in 2015

While IBM was building their colossal headquarters at nearby Havant they needed a short-term base for 750 employees. Their architect, Philip Dowson of Arup Associates, recommended Foster Associates for the temporary building. IBM wanted an off-the-peg solution that could be erected quickly and cheaply. Foster argued that for the same cost IBM could have a bespoke building of high architectural and environmental standards.

Whereas at Havant, component functions are split into separate buildings, Cosham houses multiple uses (offices, computers, a restaurant and communications centre) within a flexible, deep-plan volume – an idea that goes back to Foster's 1967 competition entry for Newport School. Tony Hunt adapted a proprietary Metsec steel frame into a self-bracing structure, with the roof void and hollow columns used for servicing. For the frameless glass wall, Foster imported the 'Pittco T-wall' system manufactured by the Pittsburgh Plate Glass company (PPG).

As Richard Einzig's 1971 photographs illustrate, the bronzed glass walls hover ambiguously between transparency and reflectivity. Inside, the open plan with its perimeter circulation ensured that every grade of employee could enjoy views out. Notwithstanding the unsympathetic addition of a pedimented porch for HMRC (which Foster likened to a car crashing into the building), Cosham has proven its intrinsic flexibility by adapting to changing needs and users over the years.

Pages 142–3: RSHP offices, The Leadenhall Building, City of London, 2007–14, Rogers Stirk Harbour + Partners.

OLIVETTI TRAINING CENTRE

Branksome Place, Hindhead Road, Haslemere, Surrey
1971–2; James Stirling and Partner
Structural engineer: Felix Samuely & Partners
Listed at grade II* in 1997

What did Eva Jiřičná mean when she described James Stirling as 'the father of high-tech architecture'?[76] Her generation were awed by the boldness and presence of Big Jim's work, much of it realized with the help of Frank Newby of Felix Samuely & Partners. After Olivetti bought Branksome Hilders, an Edwardian house by E J May, it was converted by Ted Cullinan and extended by Stirling. The new wing resembles a dismantled machine. A strip of double-banked classrooms is snapped in two; at the break is a multi-function hall and tapered link to the house. An enjoyment of mechanisms tells in the cleaning gantry, designed to splay across the glazed link, and the motorized partitions of the hall, which rise to a cruciform fly tower.

The wings are clad in all-over GRP panels, whose colour scheme morphed from Automobile Association yellow to lime and lilac, and finally to planners' cream and parchment.[77] The moulds were handed over to Olivetti so that the teaching wings could be further extruded; 'I don't give a damn whether the building is three times as long,' said Stirling, 'you can chop and change the candy bar.'[78] A singular work for an international corporate patron, Stirling's extension takes up the theme of architecture as product design, one influence being Mario Bellini's Divisumma 18 calculator for Olivetti.

BUSH LANE HOUSE

80 Cannon Street, City of London
1972–6; architects and engineers Arup Associates
(project architect Dick Raines)

When the Centre Pompidou opened in January 1977, the ten-storey Bush Lane House was a point of reference. Both buildings incorporate Arup's expertise, specifically fire protection by water cooling. In the case of Bush Lane House, an external lattice frame of tubular stainless steel is filled with water (including anti-freeze and a rust inhibitor). That provides 30 minutes of fire resistance; another half hour is gained by projecting the frame 0.6m (2ft) clear of the curtain wall. The substructure and foundations were designed to sidestep the planned tunnels of the Fleet Underground line (the programme was later scrapped). The ground floor was kept open as it was required by London Transport; it was later infilled with a glass box designed by Arup Associates.

While Bush Lane House stands within Arup's engineering tradition, it partakes of the space-age fantasy of Anne Tyng and Louis Kahn's Philadelphia City Tower scheme (1952) and Archigram's Plug-in City (1962–4). The design team went outside the building industry to find the right products and technologies. Stainless-steel spun castings came from a firm that supplied the brewing and dairy industries, while the soffit panels were supplied by a German aircraft manufacturer. The rounded, oversized nodes and vents recall contemporary product design; bespoke graphics were designed by Pentagram. The design language of the external frame is taken up at the adjacent Cannon Place (2007–11, Foggo Associates), which straddles the mainline station.

beauty
pharmacy
checkout
welcome

WILLIS FABER & DUMAS

Willis Building, 16 Friars Street, Ipswich, Suffolk
1973–4; Foster Associates
Structural engineer: Anthony Hunt Associates
Listed at grade I in 1991

Foster Associates' headquarters for the insurance broker Willis Faber & Dumas combined technical innovation with social facilities and an urban sensibility. As site boundaries were subject to revision during the design phase, Foster chose a low, deep-plan building that filled its site like an egg cracked into a frying pan. This created an undulating perimeter wall that could be tweaked during the design process. It is sheathed in frameless bronzed glass, creating an enigmatic presence that refracts the Ipswich townscape before, by night, dissolving into an illuminated interior. A full-height atrium projects through three storeys, with a pair of escalators silently ascending to a roof-top restaurant and roof garden. The atrium and four service cores slot into a 14m-square (46ft) column grid, supplemented by a 'necklace' of perimeter columns. Services are routed through a suspended ceiling and an early iteration of a raised access floor.

Perhaps the boldest move was the suspended glass wall assembly developed by Martin Francis with Pilkington's. Each 2 × 2.5m (6.5 × 8ft) sheet is attached to its upper or lower neighbour with a brass patch plate; a hinged connection copes with angled junctions. Each chain of glass is hung from a continuous top rail by a single bolt. Glass fins are attached independently to restrain the wall against lateral wind loads. In 1991 Willis Faber became the first high-tech building to be listed, prompted by proposals to infill the ground-floor swimming pool. A set of management guidelines were later agreed that streamlined the process of listed building consent.

Further reading:

Gabriele Bramante, 1993, *Willis Faber & Dumas Building: Foster Associates*. London: Phaidon.

BRIARCLIFF HOUSE

99 Eastmead, Farnborough, Hampshire
1981–3; Arup Associates (Group 5)

In the 1970s the multi-disciplinary teams at Arup Associates turned out a sequence of disciplined, refined corporate offices in landscaped surroundings. At CEGB Bristol, Lloyd's at Chatham, Gateway House at Basingstoke and the IBM campus at Portsmouth, structure, services and space are resolved into a multi-layered 'tartan grid'. At Briarcliff House that expertise was applied to a speculative brief and an urban site, anticipating the practice's Finsbury Avenue and Broadgate developments in London. It is a U-shaped building of five storeys, surrounded on three sides by a distributor road. At street level, a pedestrian mall leads to the Kingsmead Shopping Centre.

The noisy, polluted and south-facing site justified a sealed and air-conditioned building with a double skin. The outer face is bronzed, frameless glazing, which cascades over a mansard plant room before skirting out to form a glazed canopy over the pavement. Along with the **Renault Distribution Centre**, it was one of the first installations of Pilkington's Planar glazing system. An inner layer of interchangeable solid or glazed panels is set back to form a 1m-deep (3.3ft) perimeter zone that accommodates mechanized sunscreens, maintenance walkways and jumbo ducts. This cavity was intended to pre-heat the air supply to the air conditioning system while providing an acoustic barrier. The concertina geometry of the planning grid suggested a structure of pre-stressed ribs spanning between perimeter ring beams. Cabling and servicing are housed within a raised access floor. Within the U is an elevated garden and restaurant pavilion – Farnborough's answer to James Stirling's Florey Building at Oxford.

nationwide
Santander
Santander
Iceland

LLOYD'S BUILDING

1 Lime Street, City of London
1981–6; Richard Rogers Partnership
Structural and services engineer: Ove Arup & Partners
Listed at grade I in 2011

Lloyd's is a paradoxical place. How could a risk-averse institution end up with a radical headquarters? How can a lucid, diagrammatic plan generate a turbulent, romantic exterior? How is flexibility reconciled with monumentality? Edward Lloyd's coffee house became a hub for maritime business from the 1680s. In 1773, 'The Society of Lloyd's' took rooms at the Royal Exchange until 1928, when it moved into a classical headquarters by Sir Edwin Cooper. In 1952–7 it was extended across Lime Street by Cooper's successor Terence Heysham. When, in the 1970s, Lloyd's contemplated rebuilding, they settled on three principles: continuity of trading, the perpetuation of a single trading space (known as The Room), and the potential for further growth and change. From a shortlist that included Foster Associates, Arup Associates and I M Pei, RRP was selected in April 1978.

Partners in charge Richard Rogers, John Young and Marco Goldschmied developed a design that could be described as Gothic insofar as the exterior is worked hard to preserve an unfettered, spacious interior. Like the Centre Pompidou, it nods to Louis Kahn's concept of served and servant spaces. Service risers, vertical circulation and toilets are deployed to six satellite towers outside the main envelope for ease of replacement. This 'inside-out' approach decongests the high-density office floors and – unlike Pompidou – permits a full-height atrium. The resulting layout – a doughnut of floorplates, serviced from the perimeter – owes something to Frank Lloyd Wright's Larkin Building (1904–6). The structure, finely honed in fair-faced concrete, is also externalized, with rings of columns girdling the atrium. After concerns about fire resistance, the cladding was switched from aluminium to stainless-steel panels, precision-made by Josef Gartner of Germany. Rolled, textured glass inspired by Pierre Chareau's Maison de Verre combined transparency with sparkle. The finishes of stainless steel, glass and concrete make up a sharp, sober palette, highlighted by careful lighting.

Chaucer
HISCOX

Due to 'rights to light' the building steps down towards Leadenhall Market, opening up the atrium whose barrel-vaulted roof channels Paxton's Crystal Palace. Crammed into the City's medieval street pattern, and now hemmed in by tall buildings, Lloyd's was never intended to be taken in at a single glance but is encountered in scattered, oblique views. It is an active, layered exterior, packed with incident and contrast but lacking the hierarchical order of, say, Foster's Hongkong & Shanghai Bank Headquarters. Early on, Rogers had sketched the curtain walls and interval towers of medieval castles. Viewed from the pavement, Lloyd's has a similarly defensive character: its ground floor is raised over an undercroft, with stairs and walkways bridging a cobbled 'moat'.

For all its flexibility, the interior is a *Gesamtkunstwerk*, boasting the City's finest interior since Soane's Bank of England. Each component stage-whispers its functions and connections. The Underwriting Room is placed on the upper-ground floor, overlooked by galleries that open directly onto the atrium. It is an intense, inward-looking and ominous space, animated by a bank of escalators. The Lutine Bell, rung to convey news of ships, forms a quasi-liturgical focal point. Other floors incorporated interiors salvaged from the Heysham Building, chiefly a reconstructed Robert Adam drawing room (originating at Bowood House in Wiltshire). In a two-year collaboration with RRP, Eva Jiřičná updated the traditional underwriters' box with a prefabricated kit of parts and designed the restaurant, the Captain's Room (now removed).

Lloyd's was always going to be an energy-intensive building (it currently has a modest EPC rating of E), so Arup's developed a dynamic servicing strategy that recycled waste energy. Luminaires in the ceiling coffers incorporate lighting, sprinklers and air-extraction units. Stale air is passed through the cavity in the triple-glazing to provide additional insulation before its heat is stored in basement sprinkler tanks. The concrete superstructure acts as a giant heat sink, soaking up heat during the day before releasing it overnight. In 1978 fewer than 4 per cent of underwriters used computers yet the globalization of finance suggested a move from face-to-face to electronic trading. The services specification was belatedly upgraded in 1981 with additional cabling, power and cooling capacity accommodated in a raised access floor.

Lloyd's was grade-I listed in 2011, highlighting the balance between flexibility and architectural integrity. Although it was designed to accommodate a flexible mix of underwriting space and lettable offices, the headquarters are no longer owner-occupied, the freehold having been sold in 1996 on a leaseback basis. With further refurbishments contemplated, Lloyd's future will surely prove as dynamic as its designer anticipated.

Further reading:

Kenneth Powell, 1994, *Lloyd's Building: Richard Rogers Partnership*. London: Phaidon.

HOPKINS PRACTICE OFFICE

27 Broadley Terrace, Marylebone, London
1983–4, 1993–4 and later; Michael Hopkins & Partners
Structural engineers: Anthony Hunt Associates (1983–4);
Ian Duncan (1993–4)

By the early 1980s Michael and Patty Hopkins had outgrown their initial base in Hampstead. They acquired a single-storey concrete structure in Marylebone – probably a surface air-raid shelter built during the winter of 1939–40. They gradually nibbled away at the triangular site, adding freestanding pavilions as the practice expanded. At the front of the site went Building 1, a double-height box that employed components from the **Patera** system. In place of Patera's three-pinned arches were substituted a tubular column and Pratt truss arrangement, still outside the envelope. There was sufficient headroom for at first one then a pair of mezzanine floors to be slotted in alongside a top-lit aisle, with elegant spiral stairs based upon those of the **Hopkins House**.

A few years later, the remainder of the inherited building was demolished, leaving only the rear tip, which was converted to a model shop, print room and kitchen. Overspill studio space took the form of an original Patera building, shipped from Barrow-in-Furness and provided with a side entrance. Behind the front gate a glass reception box was installed, based on the display cases the practice designed for the Victoria and Albert Museum. The parts are linked by a covered way with a tensile fabric canopy – by then a Hopkins signature. In 2024 Hopkins Architects celebrated four decades at Broadley Terrace, which has itself become a showcase of the practice's work.

SOLID STATE LOGIC

25 Spring Hill Rd, Begbroke, Oxford
1986–8; Michael Hopkins & Partners
Structural and services engineer: Buro Happold

Founded by Colin Sanders in 1969, Solid State Logic (SSL) made the mixing consoles that, installed at Abbey Road Studios, Virgin Records' Townhouse Studios and The Power Station in New York City, defined the punchy, shiny sound of 1980s pop. Sanders' firm was high-tech in the original sense yet Hopkins' design suggests a shift away from a technological aesthetic.

Situated in the grounds of a late 19th-century house, it is a square building on two levels linked by an atrium, a little like Mies van der Rohe's Bacardi building. This brought together manufacture, assembly, and research and development functions, with administrative offices staying in the old house. The upper floor is jettied out and its façade is layered with full-height sliding glass doors, external venetian blinds and angled fire-escape stairs. Tubular steel columns support an in-situ concrete waffle slab with a tapered edge. It is finished in shot-blasted white concrete, with domed coffers in the soffit housing lighting and ventilation units. Above is a space-frame roof structure that is – uncharacteristically for Hopkins – concealed by a metal suspended ceiling. The atrium incorporates environmental controls including perforated sun louvres and domed rooflights with mechanized lids controlled by a thermostat.

SSL points the way to a Miesian strain in Hopkins' oeuvre, which includes 1–3 New Square, Bedfont Lakes, and the reception building at **Schlumberger** (both of 1990–2). It makes an interesting contrast with the Cotswolds Po-Mo of Solid State's earlier R&D centre at Stonesfield (1981, Doug Clelland with Eric Parry).

5 LONGWALK ROAD AND 3 FURZEGROUND WAY

Stockley Park, Hayes, Uxbridge, London
5 Longwalk Road, 1987–9; Foster Associates
3 Furzeground Way, 1989–90; Ian Ritchie Architects
Structural and services engineer: Ove Arup & Partners
Stockley Park landscape included at grade II
on the Register of Parks and Gardens in 2020

Stuart Lipton came to specialize in developing sites that no one else would take on. While Finsbury Avenue Square and Broadgate were rising from a redundant British Rail site next to Liverpool Street Station, he was terraforming 350 acres of landfill into a business park ten minutes from Heathrow. A 1985 masterplan by Michael Lowe of Arup Associates set out a strategy and a series of ground rules. Over 3 million cubic metres (106 million cubic feet) of remediated landfill were removed to the north of the site to terraform a country park and golf course in the green belt. The cleared site became the business park, with a landscape of informal lakes and semi-mature trees designed by Bernard Ede with Charles Funke. The image of white-clad pavilions within a leafy campus derived from West Coast science parks.

Lipton demanded a fast-track, low-cost programme that used prefabrication and construction management techniques to ensure rapid completion of the building's structure and services (the so-called 'shell and core'). Much was predetermined, including steel frames and floor layouts, with their skinny atrium; the fit-outs were completed by tenants. As at Broadgate, the building envelope became the main vehicle for architectural expression. Ritchie developed a double-glazed version of the Planar glazing system with Pilkington, applying low-emissivity coatings and a white ceramic frit (a technique developed for oven doors). Curved stainless steel solar shades are mounted on a slender, pin-jointed pergola. Inverting Lipton's 'pitched roof rule', Foster devised three staggered gull-wing roofs carried on Y-shaped columns. As at the **Renault Distribution Centre**, the central entrance is recessed behind the frame, creating a full-height portico.

EXCHANGE HOUSE

12 Exchange Square, Broadgate, City of London
1988–90; architects and engineers Skidmore, Owings & Merrill

Part building, part bridge, Exchange House is an 11-storey 'air rights' building that spans the tracks approaching Liverpool Street Station. Its externally expressed parabolic arches convey a brute strength at odds with the intricate, craft character of British high-tech. The project was among the most technically challenging elements of SOM'S masterplan for Stuart Lipton's Broadgate development. Like the Chestnut-De Witt Building (1964–6), the John Hancock Center (1965–70) and the Sears Tower (1970–4), Exchange House is the fruit of a creative dialogue between the architects and engineers of SOM'S Chicago office, in this case Bruce Graham and Srinivasa 'Hal' Iyengar.

Graham long dreamed of an exposed structural steel frame. What made it possible was the analytical approach of 'fire engineering'. At Exchange House, this determined a strategy based on intumescent coatings, active fire-protection measures and setting the frame 1.5m (5ft) outside the curtain wall. The design team initially considered a primary structure of multiple downward chevrons or a suspended catenary arch before opting for parabolas. There is a total of four segmented, tied arches; two on each elevation and a further pair traversing the interior, piercing a pair of staggered light wells.[79] Steel trusses bear upon six massive granite-faced piers. The transfer structure opens up the ground floor to views from Primrose Street to Exchange Square and the arched train shed beyond.

IMAGINATION LONDON

25 Store Street, South Crescent, Bloomsbury, London
1988–9; Herron Associates
Structural engineer: Buro Happold

With this intervention Ron Herron returned to the notion of a 'quiet technology', developed in his 1968 Archigram project Tuned Suburb. 'It took the thesis that it was possible to put into the existing environment a new, strange, slightly magical architecture that somehow could exist shoulder to shoulder.'[80] His client, Imagination, was founded in 1978 by Gary Withers, a pioneer of the 'brand experience' through spectacular events and exhibitions. A six-storey building fronts the crescent (a hostel of 1906 for the employees of the department store Bourne & Hollingsworth) with a lower mews terrace behind.

Herron immediately saw the potential for enveloping the intervening lightwells with 'a very lightweight, transparent structure counterpointed by the two solid walls with holes punched through from top to bottom'. At first he considered a glass roof. Then came the idea of 'throwing a fabric roof over it and wrapping it up like a Christo'.[81] In collaboration with Ian Liddell of Buro Happold, the project architect John Randle developed a tensile membrane structure of translucent PVC-coated polyester, pre-stressed by aluminium umbrellas on stainless-steel tie rods. Herron's initial idea of carrying the fabric down the sides was disallowed by fire officers. The atrium is bridged at each level with gossamer walkways on pin-joined, triangulated trusses. The concept, the structure even, is simple enough, yet the top-lit space is more than the sum of its parts. It has, in Sutherland Lyall's words, 'a kind of mysterious otherworldliness which is not easy to explain'.[82]

Further reading:

Sutherland Lyall, 1992, *Imagination Headquarters: Herron Associates*. London: Phaidon.

BRACKEN HOUSE

1 Friday Street, City of London
Original building 1955–8; Richardson and Houfe
Converted and altered 1989–91; Michael Hopkins & Partners
Structural and services engineers (1989–91): Ove Arup & Partners
Listed at grade II in 1987; upgraded to grade II*,
including Hopkins phase, in 2013

Designed by Albert Richardson, Bracken House was completed in 1958 as the headquarters of the *Financial Times* in the reconstruction area around St Paul's Cathedral. Tall office wings, clad in pinkish brick and Hollington sandstone in homage to the FT's rosy pages, bookended a recessed printing hall. With production moving to a new **print works** in Docklands, a competition for the 'redevelopment or conversion' of Richardson's building was won by a scheme by Michael Hopkins & Partners for its replacement. A rethink became necessary in August 1987 when Bracken House became the first post-war building to be listed. Returning to one of Richardson's sources, Guarini's Palazzo Carignano in Turin, Hopkins replaced the central press with an elliptical block that negotiates the differing alignments and floor levels of the wings. The principal entrance was brought around to Friday Street, giving into a double-height entrance hall and a top-lit atrium.

Bracken House heralds high-tech's contextual turn. Richardson's cornice and attic levels are carried across but the rhythm of pilasters and recessed bays is reversed to give tiers of canted bays in glass and gunmetal. These take their cue from Liverpool's Oriel Chambers (1864, Peter Ellis) and 124 Rue Réaumur, Paris (1904, Georges Chedanne), as well as Richardson's slender bays on Cannon Street. Hopkins insisted upon load-bearing external walls, detailed to show how the weight of the floor beams is successively transferred via cast-iron nodes to colonettes, cantilevered brackets and piers of Hollington stone. In 2019 the FT returned to Bracken House, which was renovated by John Robertson Architects.

Further reading:

Colin Amery, 1992, *Bracken House*. London: Wordsearch.

83
20

INSIGNIA HOUSE

83–5 Mansell Street, Whitechapel, London
1990–1; Elana Keats & Associates, and John Winter & Associates with Jonathan Ellis-Miller
Structural engineer: Ove Arup & Partners

Insignia House is an infill development on the City fringe, providing seven floors of office space on a standard speculative brief. It was designed by Jonathan Ellis-Miller, then an associate with John Winter. Its principal claim to interest is a spectacular freestanding glass wall that showcases an 'entrance atrium'. This display of high-tech bravura was Miller's creative response to a planning stipulation that a 3m (10ft) strip of land left over from a previous road-widening scheme could not be used as office space.

The bolt-fixed glazing, detailed with help from Arup's, betrays the influence of Grimshaw's recently completed ***Financial Times*** **Print Works**. Sheets of 10mm (0.4in) armour-plate glass are supported by cleats attached to steel lugs. These are welded to two tubular steel masts, pin-jointed top and bottom, with horizontal bracing at 2m (6.5ft) centres. Steel tension wires provide bracing in the vertical plane. This encloses a void that rises the full height of the building to a space frame roof. Behind are the exposed downstand beams of the office floorplates. The spectacular glass front belies the modest scale of the development and adds a touch of drama to Mansell Street, especially when floodlit. The calmer rear elevation on West Tenter Street is an all-glass curtain wall with a minimal steel frame, with the upper two storeys set back.

WESTERN MORNING NEWS

17 Brest Road, Derriford, Plymouth, Devon
1991–3; Nicholas Grimshaw & Partners
Structural engineer: Ove Arup & Partners
Listed at grade II* in July 2015

The building was commissioned after the Western Morning News Company, part of Northcliffe Newspapers, planned to move out of overcrowded city-centre offices and invest in new printing presses. The managing director, Jerry Ramsden, had been impressed by the *Financial Times* **Print Works** in London's Docklands and sought 'a landmark to show off to the world'.[83] Ramsden certainly got his landmark, a landlocked (one-)liner with a fully glazed hull. The nautical allusions are more literal than, say, Ralph Erskine's contemporaneous Ark at the Hammersmith Flyover, and sail close to post-modern symbolism. Unlike most modern print-media operations, the *Western Morning News* integrated editorial and advertising offices with production facilities. Partly for this reason, it departs from Grimshaw's in-the-round ethos, having a prestige bow and a functional stern.

Boarding the building via a gangplank, visitors arrived at a triangular atrium at the heart of the office section. At the top, a 22m-tall (72ft) bridge houses a boardroom and observation gallery with views of Plymouth Sound. From an external structure of steel 'tusks' are hung 700 tapered and facetted panels of bolt-fixed glazing, fixed with glazing arms of ductile iron. The leaning glass wall reprises a similar detail from Grimshaw's 1989 competition entry for the Bibliothèque Nationale de France; here printing presses rather than book-retrieval machinery were put on show. Since the *Western Morning News* moved out in 2013 the building has been let to commercial tenants.

CHANNEL FOUR TELEVISION

124–6 Horseferry Road, Westminster, London
1992–4; Richard Rogers Partnership
Structural and services engineer: Ove Arup & Partners
Listed at grade II in 2023

Channel 4 was launched on 2 November 1982 as a publicly owned, commercially funded public service broadcaster with a remit to make innovative, experimental and distinctive programmes. As its audience share increased and the switch to digital broadcasting loomed, the channel's chief executive Michael Grade and chairman Sir Richard Attenborough chose to build a headquarters. A Westminster site was found and a limited competition was held in late 1990, narrowed down to Stirling Wilford, Bennetts Associates and RRP. Channel 4 was founded as a commissioner and a broadcaster but not a producer of programmes, so their main requirements were for offices and prestigious spaces to receive clients.

RRP designed a concrete-framed metal and glass structure 'in the **Lloyd's** mould', as John Young put it.[84] It has an L-plan of office wings, hinged by a corner entrance. This takes the form of a concave glass curtain wall, supported by a cable-net structure devised by Peter Rice (and recently refurbished by Arup). On one side are conference rooms stacked on pin-jointed frames; on the other are wall-climber lifts, service elements and a transmission tower in C4's scarlet livery. Constructivist clusters of this sort featured in several Rogers schemes of the 1980s, including their 1982 National Gallery competition entry. Unlike Foster Associates' ITN Headquarters near King's Cross (1987–90), Channel 4 was wholly owner-occupied, although the office space was specified so that the building could readily be sold or tenanted. In January 2024, the broadcaster announced their intention to relocate as part of a restructuring.

Further reading:

Deyan Sudjic, 1996, *Channel 4 Headquarters, London*. London: Blueprint Media.

124 HORSEFERRY ROAD

INLAND REVENUE

Castle Meadow Road, Nottingham
1993–5; Michael Hopkins & Partners
Structural and services engineer: Ove Arup & Partners
Listed at grade II in 2023

In 1989 the Inland Revenue decided to relocate 2,000 civil servants to Nottingham as part of a government programme to create jobs in unemployment blackspots. A central site was identified on former railway land, adjoining Nottingham Canal and having fine views to the castle. After the city council rejected design-and-build proposals for such a sensitive site, an open competition attracted 134 entries from which Michael Hopkins & Partners' was selected. Six blocks line a gently curving, tree-lined road, crossed by secondary routes. The 'perimeter plan' allows a comparatively high-density, low-rise complex as well as a return towards shallow office floors for natural light and ventilation. It had the additional benefit of permitting a phased development strategy, while the blocks could be leased if plans changed.

As at **Portcullis House**, load-bearing brick piers support curved concrete floorplates, both elements prefabricated off-site to speed up construction. Their combination suggests Louis Kahn's legible façades as well as the red brick and jack arches of Nottingham's warehouses. The masonry elevations are shaded by a cantilevered attic storey and punctuated by tall corner towers of glass block and steel. A central amenities building reprises the tensile fabric structure of **Schlumberger**. The glazed stair towers act as 'solar chimneys', using the sun's warmth to draw stale air from the office floors. After HMRC vacated the complex in 2022, Nottingham University outlined proposals to convert it to a campus for their final-year and post-graduate students. It will be their second Hopkins commission, following their 1996-9 masterplan for the Jubilee Campus, north of the city.

88 WOOD STREET

City of London
1995–9; Richard Rogers Partnership
Structural and services engineer: Ove Arup & Partners

88 Wood Street is a high-stakes site, with the Wren tower of St Alban's Church, McMorran and Whitby's Police Station and Terry Farrell's Alban Gate within striking distance. The project began as a European headquarters for Daiwa Securities, for whom RRP designed a curved atrium building that stepped up to 12 storeys on London Wall. It gained planning permission but fell foul of the early 1990s recession and the listing of a telephone exchange on the site. By 1994 Daiwa wanted speculative offices on a reduced budget, providing a third more accommodation without exceeding the building envelope of the previous scheme.

RRP'S quietly efficient redesign is based on a trio of linked slabs of 10, 14 and 18 storeys, a tactic popularized by Foster Associates' Hong Kong Bank headquarters. This respected the cornice line to Wood Street while allowing the building to be flexibly subdivided into separate lettable areas. Pre-stressing and post-tensioning the concrete floorplates slimmed them enough that an additional storey could be squeezed in. The storey-height glass panes of 3 × 4m (13ft) are clasped by frameless fixings, with giant cross-braces to the end bays. Like **Lloyd's of London**, Wood Street features a triple-glazed 'climate façade' in which exhaust air is drawn through a 140mm (5.5in) cavity into extract ducts in the ceiling. The same gap is fitted with perforated venetian blinds operated by photocells on the roof. Access and service towers are housed within smartly glazed enclosures. Self-quotations of red and blue ventilation funnels and bright yellow steelwork provide playful accents. In 2021 No. 88 was refurbished by Morrow + Lorraine.

LLOYD'S REGISTER

71 Fenchurch Street, City of London
1996–2000; Richard Rogers Partnership
Structural engineer: Anthony Hunt Associates
Services engineer: Ove Arup & Partners

Lloyd's Register (LR) shares the same origins as Lloyd's but has long been a separate organization. In 1901 the firm moved into a splendid headquarters by T E Collcutt at 71 Fenchurch Street. By the early 1990s LR contemplated moving out to a greenfield campus at Liphook, Hampshire, for which Graham Stirk prepared designs. In the event, he was asked to reconfigure their base in the Square Mile, creating an architectural vitrine concealed behind retained buildings (or, in the case of Coronation House, propped façades). Two slender slabs of 14 and 11 storeys and a seven-storey block behind Lloyd's Avenue are sandwiched by glazed atria. The stack of vertical elements fans out on a radiating grid to resolve the wedge-shaped site.

The towers are fully glazed with motorized louvres and bright yellow blinds controlled by rooftop photovoltaic cells. Services are flaunted in the **Lloyd's** tradition, with glass wall-climber lifts and stairs to the fore and toilets, goods and service risers behind. The service cores do not stiffen the structure, so raking 'K-braces' were added to the concrete frame. Inside, ceiling-mounted service units (known as passive chilled beams) are integrated into the exposed concrete vaults of the office floors. Behind Fenchurch Street, the former churchyard of St Katherine Coleman (the church having been swept away in 1926) was laid out as a public space. Successive proposals for a gateway building replacing the Edwardian Nos 68–70 were turned down by City planners. So it is only after stepping through its arched passageway that one is confronted with RRP'S *tour de force*.

Further reading:

Nigel Watson, 2010, *Lloyd's Register: 250 Years of Service*. London: Lloyd's Register.

EAST INDIA ARMS
Nº 67
EAST INDIA ARMS
Shepherd Neame ~ Britain's Oldest Brewer
EAT.

SWISS RE BUILDING

30 St Mary Axe, City of London
2001–3; Foster + Partners
Structural engineer: Arup

Swiss Re ushered in a new City cluster of skyscrapers that, a quarter-century on, threatens to eclipse it. The opportunity to rebuild arose after the Baltic Exchange was seriously damaged by an IRA bomb blast in April 1992. At 386m (1,266ft) in height, Foster's pathfinder, the Millennium Tower, was rejected by City planners and regulatory bodies as out-of-scale. The scheme is remembered for the *Guardian*'s sobriquet, 'erotic gherkin'.[85] Several organic forms ('loaf of bread', 'egg on a podium') were thrown out before a fat pinecone was arrived at. In successive tweaks this became slimmer and taller, reaching 180m (590ft), 2m (6.5ft) more than the structure's circumference at its widest point and 3m (10ft) shorter than Tower 42. Circulation and services are carried in a central core circumscribed by a steel frame, taking out a big chunk of the gross floor area.

A perimeter diagrid provides stiffness and, thanks to its aerodynamic profile, wind resistance. Disappointingly, the diagrid had to be fireproofed and encased in aluminium packaging. By cutting away wedges from the floorplate the result was a series of rectangular fingers radiating from the core like a six-armed starfish. Each successive storey is rotated clockwise by 5 degrees, an ingenious twist that allows the triangular lightwells to spiral up the diagrid, their helical path marked in tinted glass. Rather like Bruno Taut's 1914 Glass Pavilion, the curved skin is resolved into facets. So on sunny days the Gherkin becomes a giant mirror ball, animating its neighbours with dancing slivers of light.

Further reading:

Kenneth Powell, 2006, *30 St Mary Axe: a Tower for London*. London: Merrell.

LLOYDS BANK
LLOYDS BANK

THE LEADENHALL BUILDING

122 Leadenhall Street, City of London
2007–14; Rogers Stirk Harbour + Partners
Structural and services engineer: Arup

The Leadenhall Building possesses a simplicity of outline that contrasts with **Lloyd's** across the road, yet it upholds the high-tech conventions of segregated services and celebrating process and movement. Its wedge profile was a creative response to the planning controls known as 'St Paul's Heights', with the tower cleaved at 10.6 degrees to preserve views of the cathedral dome from Fleet Street. It was duly nicknamed 'The Cheesegrater' by City planner Peter Rees.

Circulation, toilets and service risers are placed on the vertical north wall, where banks of speedy lifts animate the view from Undershaft. Its braced perimeter frame can be classified as a trussed tube, in the tradition of SOM's Hancock Center in Chicago. Leadenhall's 'megaframe' is sandwiched within a double-skin façade and divided into eight braced modules, each of seven storeys. A secondary structure of K-braces provides further stability at the north corners. Some 85 per cent of the components were fabricated off-site, including the nodes that connect up to six members, distributing loads of up to 6,000 tonnes.

Like **Swiss Re**, attention is paid to the way in which the building meets the ground and the sky. Generators, boilers and cooling towers are distributed across four attic storeys, while the pavement gives onto a semi-enclosed, pseudo-public space, a gesture of 'giving something back' to the City typical of Rogers urbanism but dependent on his clients' largesse. In December 2015 RSHP moved out of their Thames Wharf offices and into the 14th floor.

THE SHARD

32 London Bridge Street, Southwark, London
2009–12; Renzo Piano Building Workshop
Structural engineer: WSP Cantor Seinuk

For now the tallest building in the UK, the Shard is an inscrutable, selfish presence above London Bridge Station. It is a deliberately unresolved gathering of planes that rise to an open crown of splinters. Less decisive is the way the tower meets the ground, with the curtain wall hitched up to expose fat girders and a jutting canopy, necessary to shield pedestrians from downdrafts. Enabling technology is not flaunted but works efficiently behind the curtain, including a slip-cast concrete core and a triple-skinned façade with a ventilated cavity and motorized blinds. But to what end? The development, by the entrepreneur Irvine Sellar, is a transparent expression of the migration of global finance south of the Thames – boosting local levels of inequality and gentrification.

It is a mixed-use programme, although one in which the functional elements – offices, restaurants, a Shangri-La hotel, luxury flats – go unstated. At the top is an observation gallery to which public access, negotiated as part of a Section 106 planning obligation, comes at a cost of £28.50 per person (champagne packages available). Of greater public benefit was the agreement to fund the rebuilding of the station concourse. (In 2012–18 the remainder of London Bridge was redeveloped by Grimshaw to a masterplan by T P Bennett and Alan Baxter Associates.) Piano's trajectory from Young Turk to oligarchitect mirrors that of his former partner: Rogers' Neo Bankside (2009–12), at the other end of Southwark Street, was criticized for 'push[ing] the role of architecture as a spectacularly effective lubricant, allowing its developers [...] to slip out of the council's planning obligations with unparalleled ease'.[86]

Further reading:

Howard Watson, 2017, *The Shard: the Vision of Irvine Sellar*. London: Constable.

SECOND HOME LONDON FIELDS

125–7 Mare Street, Hackney, London
2018–19; Estudio Cano Lasso
Structural engineer: Webb Yates
Façade consultant: Lastra & Zorilla

For many, working life in the 21st century is more nomadic, isolated and precarious than ever before. Sam Aldenton and Rohan Silva noticed that the commercial property market hadn't kept pace with trends towards informal, flexible and remote working. In 2014 they opened Second Home Spitalfields, a co-working space boasting a nursery, café, rooftop bar and cultural programme. Further locations followed in Lisbon and Holland Park (occupying Richard Rogers' former studio), designed by SelgasCano of Madrid. At Mare Street, sister practice Estudio Cano Lasso retrofitted Triangle House, a 1960s extension to Hackney Technical College. The interiors were kitted out with a fidgety, Instagram-friendly array of potted plants, pods and partitions, with peepholes between floors.

Lead architect Gonzalo Cano Pintos considered several façade interventions, including a lightweight scaffolding 'full of mirrors, vegetation, banners', and a lightweight metal mesh based on the **Snowdon Aviary**.[87] Seeking the translucency of a cloud or a soap bubble, he settled on a twin-skin membrane of ETFE plastic, detailed by Webb Yates with circular cut-outs that ventilate the façade and reveal the supporting armature. It's a superficial gesture: the 1960s window-wall behind was stripped and overglazed. As influences Pintos cited the tensile architecture of Frei Otto and the veil worn out of modesty or vanity by the Roman empress Poppaea Sabina. As spontaneous as a pop-up, the premises proved about as transient. Having been hit by the pandemic, it was vacated by Second Home in 2023 and at the time of writing is available for rent.

Kids around
Kids around
HOLLISTER

6 | retail

SAINSBURY'S CANTERBURY

Kingsmead Road, Canterbury, Kent
1983–4; Ahrends Burton and Koralek
Structural consultant: Anthony Hunt & Partners; executive engineers: Ernest Green & Partners

Architect-designed supermarkets could be classified into 'decorated sheds', where design input was limited to what John Outram described as the 'Façade to the Carparking', and those rarer situations in which the architect was permitted to tackle the shell.[88] Sainsbury's tried the former at Canley, near Coventry (1994; Lifschutz Davidson) and Marsh Mills outside Plymouth (1991–4; Dixon Jones, with Peter Rice). The latter includes the Brentford Homebase (1987; Nicholas Grimshaw & Partners) and the eco-Sainsbury's at Greenwich (1999, demolished 2016; Chetwood Associates).

ABK's design for the Canterbury branch, developed with Tony Hunt, won a limited design competition in 1982 (assessors included Roy Strong and Norman Foster). Its high-tech image, pointing to the spires and pinnacles of the cathedral, was a notable departure from the pseudo-vernacular mainstream of supermarket design. A stepped plan gives a 36 × 65m (120 × 215ft) clear-span retail space, warehouse and unloading bay, each separated by a 3m (10ft) service strip. The roof structure is suspended from paired white tubular-steel masts and outriggers with tension rods, while a fabric canopy is rigged along the front. Cladding is of aluminium sandwich panels with blockwork to the service zones. An upper cranked section acts as a slip joint to take up differential expansion between roof and wall. ABK had no say in the fit-out of the interior, which was kept in-house. In the late 1990s the supermarket was extended to the west with a new entrance in a sympathetic idiom.

Pages 190–1: Ashford Designer Outlet, Kent, 1999–2000, Richard Rogers Partnership.

Sainsbury's
Sainsbury's

SAINSBURY'S CAMDEN

Camden Road, London
1986–8; Nicholas Grimshaw & Partners
Structural engineer: Kenchington Little & Partners
Listed at grade II in 2019

In July 2019, Camden's Sainsbury's became the UK'S first listed supermarket. Self-service retail took off after the Second World War as cars and refrigerators became more affordable. As supermarkets and their car parks expanded it became trickier to fit them onto urban sites, prompting a shift to shopping centres and peripheral locations. This central London branch bucked the trend, as it formed part of a mixed-use development that included workshops and speculative housing (**Grand Union Walk**). NG&P were appointed on the advice of a vetting committee chaired by Colin Amery, who reported directly to the firm's chairman, Sir John Sainsbury.

From Camden Road, the store is entered via a low vestibule from which travelators take shoppers and their trolleys down to a basement car park. Ancillary functions (plant, offices and staff facilities) are raised onto first-floor strips running along the long sides. This gave a two-storey street front to match the cornice line of the listed terrace opposite. The requirement for a 43m (141ft) clear span, together with the first-floor perimeter accommodation, suggested an elaborate structure with outer cantilevers. At each end, arched roof trusses are bolted to cantilever girders. In turn the girders are anchored by clusters of tension rods, which run down to steel shoes mounted on tall concrete plinths. Grimshaw wanted to give the store an assertive, urban presence. The exteriors are based on three elements: displayed structure; a glazed shopfront, recessed and fitted with louvres; and a palette of glass, steel and aluminium. The pressed aluminium cladding panels are a development of those devised for the **Herman Miller factory**.

ASHFORD DESIGNER OUTLET

Kimberley Way, Ashford, Kent
1999–2000; Richard Rogers Partnership
Structural engineer: Buro Happold

Selling seconds or end-of-line goods at a discount, the retail outlet is the millennial offspring of the factory shop and the shopping centre. Ashford's outlet occupies the site of the Kimberley railway works, cleared prior to the construction of the adjacent Ashford International railway station on the Channel Tunnel Rail Link (rebranded High Speed 1). RRP were approached in 1996 by BAA McArthurGlen, a transatlantic venture with whom they collaborated at Heathrow Terminals 1 and 5. Instead of a mall in a sea of cars, Ashford offers a ring of shops surrounding a sunken parking area, a format trialled at McArthurGlen's first outlet, Cheshire Oaks at Ellesmere Port (1995).

Ashford's teardrop layout is squeezed between the Hastings line and an arterial road, from which it is screened by a service route and an earth bund recycled from decontamination works. Vehicle access is from the south and pedestrians enter via the northern tip (remodelled and extended in 2019 by Chapman Taylor). Needing to conjure a distinctive presence from a low budget, Buro Happold suggested fabric engineering. A continuous PVC membrane, hung from 22 raked and tapered masts, oversails the steel and glass retail units. Hangar cables slung from each mast support intermediate ridges, creating a series of triplets. Uplit at night, the diaphanous roof creates an ethereal, festive atmosphere likened by its architect to 'a Bedouin tent in an English meadow'.[89]

JACK WILLS
Barbour

2 | industry

SAPA FACTORY

Hydro Extrusion UK Ltd, Saw Pit Industrial Estate,
Tibshelf, Derbyshire
1972–3, 1977; Foster Associates
Structural engineer: Anthony Hunt Associates

Based in Vetlanda in southern Sweden, Scandinavian Aluminium Profiles AB decided that exports warranted a UK production facility. Foster was appointed on a crash schedule, with inflatable temporary warehousing and prefabricated office units installed in five days. The brief was received in July 1972, foundations were laid in September and production started the following June. The aluminium extrusion process required a span of 21m (70ft) and an 8m (26ft) clear height, determining structural bays of 21 × 100m (69 × 330ft). A total of four bays were completed in phases without disrupting production. The linear plan tracks the production sequence, with unloading and storage of ingots followed by the extrusion press, water and air quenching and run-on table, finishing with packing and dispatch.

The mute, 100m-long (330ft) exteriors are relieved only by people- and vehicle-sized openings. It is clad in Italian-made profiled sandwich panels of stove-enamelled steel. Project architect Tomm Nyhuus wanted an exterior of lime green, egg-yolk yellow or polished silver but off-white was agreed after planners threatened 'muted green or brown'.[90] To compensate, the interior became a riot of colour-coding in blue, orange, green and yellow, influenced by Craig Ellwood's Scientific Data Systems. SAPA's mission – 'to make profits and have fun doing it' – was in tune with Foster's idea of industrial democracy.[91] Employees were consulted in the design of the facility and their changing rooms and mezzanine canteen were equipped to a high standard. Foster's designed a single-storey office pavilion but in 1979 SAPA opted for an aluminium-framed block by William Saunders & Partners.

Pages 198–9: Schlumberger, Cambridge, 1983–5, Michael Hopkins & Partners.

WAREHOUSE AND SHOWROOM FOR MODERN ART GLASS

Howdens Erith, 70 Hailey Road, Eastern Industrial Estate,
Thamesmead, London
1973; Foster Associates
Structural engineer: Anthony Hunt Associates

Modern Art Glass (MAG) of Leeds were one of the UK'S largest sub-contractors of glass wall systems. On relocating their southern branch from Plumstead to a new industrial estate at Thamesmead they contacted Foster's, having been a nominated subcontractor on the Fred Olsen Operations and Amenity Centre at Millwall and **IBM Cosham**. MAG wanted a low-cost warehouse with offices and showroom. Everything was squeezed into a tube with wrap-around walls. The extrusion is sliced off with a glass curtain wall that served as an advertisement for the client's product. In place of the futuristic image of Rogers' Zip-Up House (page 16), the Thamesmead shed reinvents the portal frame and corrugated cladding of the industrial vernacular.

The architect Bill Howell thought that designers of buildings fall into one of two categories: detail refiners or detail eliminators.[92] Foster belongs to the latter camp. By substituting the eaves/gutter assembly with a curved corner, the result resembles an outsized consumer product. The showroom and mezzanine office were placed at the fully glazed end bay, separated from the warehousing by a service strip. The glass wall, designed with cladding consultant Martin Francis, consists of bronze-tinted, toughened glass carried on suspended tubular steel mullions and externally secured with steel and neoprene clamping strips. This bespoke system was developed as a trial run for **Willis Faber & Dumas** (which, in the event, incorporated a structural glass system developed by Pilkington). The building is still there, extended by two structural bays, and painted light grey instead of the vibrant blue that Foster insisted upon.

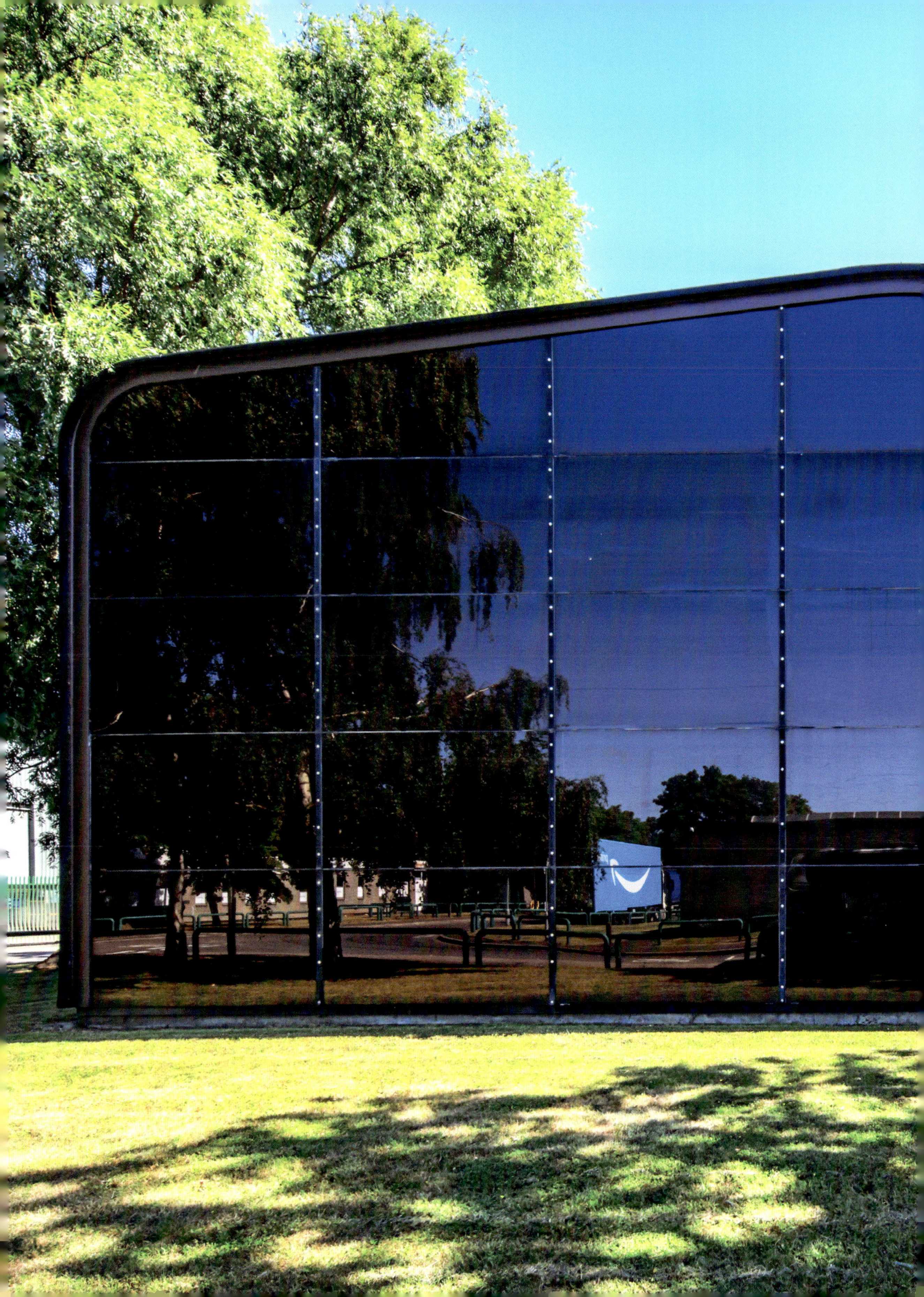

SCHREIBER FURNITURE FACTORY

Howdens Joinery, Astmoor Industrial Estate,
Astmoor Road, Runcorn, Cheshire
1973–4, 1978–9; Runcorn Development Corporation architect's department, later Brock Carmichael Associates
Structural engineer: Ewart James
of Runcorn Development Corporation

Long before the arrival of IKEA, Chaim Schreiber (1918–84) sold modern furniture for the tightest budgets, establishing a chain of furniture centres and moving into flat-pack. In 1972 he agreed to relocate to Runcorn New Town, striking a deal with its Development Corporation to erect a 42,000m^2 (450,000ft^2) factory employing 750 workers. Amid turbulent economic conditions Schreiber pulled out mid-contract. He later returned to the site and fitted out the factory to a revised design by the original architect, Peter Carmichael. It boasted a distinctive exterior, with broad eaves, tinted glazing and paired and cross-braced stanchions, painted orange (Runcorn Development Corporation's house colour).

Schreiber had two demands: the flexibility to reconfigure the production process and a decent working environment, with no segregation between blue- and white-collar workers. A linear layout follows the production line: chipboard and other raw materials arriving at the east end while cardboard cartons leave the despatch bays to the west. Offices, a display area and a canteen are housed in a first-floor section at the front. Along the long edge runs a long service spine, planned with the intention of future expansion. The roof structure is of colossal triangular lattice trusses of tubular steel. Services are threaded between the yellow-painted openwork girders, leaving a 7.5m (25ft) clear height for aerial conveyor belts and a glazed walkway. The heating system was powered by waste wood from the manufacturing process, making the plant largely self-sufficient in energy. Today fitted bedrooms and kitchens are made here by Howdens.

Further reading:

Avital Schreiber Levy, 2010, *Chaim Schreiber's Way*. Tel Aviv: Avital Schreiber Levy.

PA TECHNOLOGY

PA Consulting, Back Lane, Melbourn, Hertfordshire
Phase I: 1974–5; Piano + Rogers. Phases II and III: 1980–1, 1983–4;
Richard Rogers Partnership
Structural engineer: Felix Samuely & Partners

PA Technology was founded by Gordon Edge in 1970 as an arm of the PA Consulting Group. Edge was previously managing director of Cambridge Consultants Ltd, a pioneer of the 'Cambridge phenomenon', the high-tech cluster of research and development start-ups around the university. In 1973 PA commissioned Piano + Rogers to design a new headquarters outside Melbourn on the Cambridgeshire–Hertfordshire border. Impact on the surrounding landscape was minimized with a two-storey building, sunk into chalk embankments and accessed via cuttings. An 'activity floor', with open-plan workspace around enclosed laboratory cores, sits like a *piano nobile* above an undercroft that houses workshops, stores and car parking.

The client's dual requirements of high-security workspace and prestige showcase are reconciled through the segregation of functions and routes. Staff access their desks from six fire-escape stairs, while visitors ascend from the double-height reception area to conference rooms. Space planning is loose-fit and flexible on a 2.4m (8ft) module, with services routed through the concrete waffle floor slab. A system of lightweight, steel-framed cladding was designed with interchangeable glass and vitreous enamel sandwich panels. The double-height reception, now altered, combined glass bricks, exposed steel frame and primary colours in tribute to Pierre Chareau's Maison de Verre (1928–32). Its industrial aesthetic is completed by a pressed steel and wire staircase inspired by the scaffolders' staircase at the Centre Pompidou. Rogers was called back to design two further phases as well as the Patcenter in Princeton, New Jersey.

HERMAN MILLER FACTORY

Bath Spa University, Locksbrook Campus, Locksbrook Road, Bath
1976–7; Farrell/Grimshaw Partnership
Structural engineer: Peter Brett Associates
Listed at grade II in 2013

It is hard to imagine a better architect–client match than Nick Grimshaw and the American furniture manufacturer Herman Miller. Farrell/Grimshaw were already their customers, kitting out their practice with Eames chairs and Action Office II workstations. In 1970, Herman Miller established a manufacturing base in the UK, taking a lease on the Bath Cabinet Makers factory (1966–7, YRM) beside the River Avon. When they decided to expand across the river, Farrell/Grimshaw won the limited competition to design their first overseas plant. A free-verse brief by chairman Max De Pree pondered a working environment which 'forgives mistakes in planning [...] has flexibility, is non-precious and non-monumental'.[93]

Opened in 1977, the Bath plant was noted for its interchangeable cladding system, which Grimshaw nicknamed the 'Action Factory' after the kit-of-parts approach of Action Office.[94] Limestone-coloured GRP sandwich panels, brown-tinted glazing (fixed or louvred) and door units could be clipped into a neoprene gasket glazing system on a 1.2m (4ft) module. The material turned corners with ease, including the recessed entrance bays and sheltered sitting out areas. The GRP skin was a development of Grimshaw's factories for Rotork Controls at Bath (1971–2) and Maryland, USA, and their headquarters for Editions Van de Velde, Tours, France (1974–5). After Herman Miller commissioned Grimshaw to design a new facility in Melksham, Wiltshire, the practice also repurposed the 1970s factory for Bath Spa University's Schools of Art and Design. The panels were repaired and retrofitted to update their thermal performance, while a recessed attic storey provides space for teaching and events.

Further reading:

Jeffrey Cruikshank, 1994, *Herman Miller, Inc.: Buildings and Beliefs*. Washington, DC: American Institute of Architects Press.

CUMMINS ENGINE ASSEMBLY PLANT

M8 Business Park, Calderhead Road, Shotts, Lanarkshire
1977–83; Ahrends Burton and Koralek
Structural, mechanical and electrical engineers: Ove Arup & Partners
Listed at Category A in 2004

Under CEO J Irwin Miller, the Cummins Engine Company of Columbus, Indiana, became a noted patron of industrial architecture. The firm wanted to overhaul their site at Shotts, mid-way between Glasgow and Edinburgh and their first production plant outside the US. ABK were recommended by Kevin Roche, who designed Cummins' engine plants at Darlington (1964–5; grade II*) and their headquarters, completed in 1984. A scheme of phased expansion was worked up with the Scottish managers and US executives. Peter Ahrends shared Miller's desire to create a humane workplace, consulting the workforce and their unions.

Unlike the loose-fit Fleetguard plant in Brittany (designed by Richard Rogers for a Cummins subsidiary), the factory layout cleaves to the process flow. Four zones – receiving; machining; stores and sub-assembly; and final assembly, testing and shipping – were transected by raised walkways that extend via covered bridges to the car park. Blue- and white-collar workers shared a mezzanine canteen that overlooks the assembly line. An all-encompassing roof – an ABK trademark – integrates structure, services and top-lights, allowing conveyors and machinery to be hung from the trusses. A zigzagging perimeter of profiled aluminium and patent glazing is punctuated by inclined struts, resembling guy ropes staking down a tent. On the approach, the plant emerges from prismatic mounds formed by landscape architect James Hope from bings (spoil heaps). In 1996 Cummins moved operations to a new facility at Manston, Kent, designed by Bennetts Associates. After being listed, Shotts was adapted to host multiple businesses.

IBM GREENFORD

Green Park Way, Greenford, Ealing, London
1977–80; Foster Associates
Structural engineer: Anthony Hunt Associates
Listed at grade II in 2013

After the **Pilot Head Office** at Cosham, IBM returned to Foster Associates with an industrial site near Heathrow Airport. Their main requirement was for a distribution centre, including a warehouse, repair shop and offices. This Foster accommodated within a single envelope. Eighteen months on, he added an installation support centre where mainframe computers could be demonstrated. A double-height machine hall was linked at high level over a service road. These disparate elements were set under a uniform roofline and given a consistent treatment, including an acid-green colour scheme. As at the Fred Olsen Operations and Amenity Centre at Millwall, blue- and white-collar workers shared facilities and enjoyed the same environmental standards. 'We wanted to get away from posh and scruffy, front and back, clean and dirty, we and they divisions,' Foster told Deyan Sudjic.[95]

Although the complex was erected quickly to a modest budget, the architecture comes from the consistent use of a limited kit of parts. Industrial elements, such as the lattice roof beams and profiled sheet cladding, are juxtaposed with a sophisticated gasket glazing system developed by the Pittsburgh Plate Glass company (PPG) and previously employed for Fred Olsen and the earlier IBM commission. Refined and ordinary elements are juxtaposed in the assembly of the lift enclosure, where butt-joined glass panels are screwed directly to welded steel box sections. Despite alterations including the flooring over of the computer hall and the addition of matching bays, the complex was listed at grade II in 2013.

Swi

ADVANCE INDUSTRIAL UNITS

Chetham Court, Winwick Quay Business Park, Warrington, Cheshire
1978–9; Farrell/Grimshaw Partnership
Structural engineer: Chief engineer, Warrington New Town Development Corporation

The flexible, clip-on approach of the **Herman Miller Factory** pointed the way to a sequence of serviced sheds by Farrell/Grimshaw, including Gillingham Industrial Park (1979–80) and Castle Park, Nottingham (1980). At Warrington New Town the model was applied to a group of advanced units devised by the Development Corporation at the Winwick Quay Industrial Estate, off the newly completed M62. Low-cost, serviced starter units were laid on at many new towns to attract inwards investment. While most advance factory units, such as Kiln Farm in Milton Keynes (1974–6, Milton Keynes Development Corporation), were light industrial spaces, those at Winwick Quay provided a flexible mix of industrial, distribution or office space.

The basic shell consists of a steel frame on a 10 × 17.5m (33 × 57ft) grid, braced by a steel roof deck. Services are on a perimeter ring, brought in through special floor panels, while prefabricated toilet cubicles can be plumbed into multiple positions. A gasketed façade system (Presslock by Modern Art Glass) allowed the pattern of glazed or aluminium sandwich panels to be reconfigured to suit tenants' needs. Grimshaw had wanted aluminium cladding at Herman Miller but this was precluded by the budget. He later specified the same Alucobond panels at a BMW distribution centre in Bracknell (1980, demolished 2015). Winwick Quay anticipates a move from the pop greens and reds of Gillingham and Nottingham to the 'airstream silver' of his later works.[96] Today Winwick Quay has lost its cool sheen; its façades are grimy, over-painted and festooned with the costume jewellery of security lights and CCTV cameras.

BESPAK

Bergen Way, North Lynn Industrial Estate, King's Lynn, Norfolk
1978–80, 1985–6; Cambridge Design (Syd Furness, partner in charge)
Structural engineer: Peter Dann & Partners
Services engineer: Max Fordham & Partners

The aerosol revolutionized the post-war pharmaceutical and cosmetic industries. Bespak was founded in 1958 by Alec Schumann, making aerosol valves for such products as Soothe Tan, Hi Styler and Yardley Spray Mist. Later the firm obtained a licence to manufacture parts for the pressurized metered-dose inhalers used to treat asthma. In the 1970s Bespak moved to an industrial estate in King's Lynn and held a limited design competition for a phased extension, won in 1978 by Cambridge Design. The brief specified that the shop floor should have the same environmental standards as the offices, including good acoustics and ventilation. A courtyard separates the new structure from the existing factory, with glazed links at both ends.

The roof structure is more elaborate than the hipped exterior suggests. A 13.5m-square (45ft) grid of tubular steel columns is spanned diagonally by trusses, with secondary trusses between the apex of each bay. This umbrella roof structure supports four parallel pitched ranges, which terminate in hipped returns. The valley gutters drain into downpipes concealed within the hollow columns. By increasing the internal height to 5.5m (18ft) it was possible to accommodate a two-storey section of open-plan offices above a canteen. In 1983 Cambridge Design were invited back to design a third phase, replacing the initial factory with an exposed steel portal frame structure on an 18 × 6m (60 × 20ft) grid. This has a tied rafter roof with valley beams spanning internal columns and supporting intermediate rafters. In both phases the frame is painted mustard yellow and exposed externally against dark brown cladding panels and tinted glazing strips.

RACKING PLANT FOR GREENE KING

Off Cullum Road, Bury St Edmunds, Suffolk
1979–80; Michael Hopkins & Partners
Structural engineer: Anthony Hunt Associates

Founded in 1971, the Campaign for Real Ale led to a resurgence in cask-conditioned beer. (Unlike keg beer, cask ale isn't filtered, pasteurized and carbonated, allowing secondary fermentation to occur in the barrel.) This led Greene King, East Anglia's biggest brewer, to commission a draught beer department in which both casks and kegs were filled, stored and despatched. Hopkins' structure encloses a straightforward process flow. At the north end, empties are unloaded and transported by conveyor to be washed and filled (or racked, in brewers' parlance) by beer tanks supplied by an overhead pipeline from the Westgate brewery. Palletized casks are then transferred by fork-lift to a temperature-controlled store before being despatched onto dray lorries.

All this happens in an extruded structure redolent of the **Sainsbury Centre** or, perhaps, an industrial variant of the **Hopkins House**, having side bays of profiled steel and open ends with up-and-over doors. As the building is situated in part of the water meadows of the River Linnet, it is raised above ground level on a concrete plinth so that rising waters do not disrupt the flow of beer. The raised plinth has the additional advantage of simplifying loading and unloading, which is sheltered by a cantilevered roof structure. This was one of Hopkins' first independent commissions but its future is in doubt: in 2024, Greene King announced plans to move from its historic premises to an industrial estate on the edge of the town.

PATERA PROTOTYPE

Original location: Victoria Road, Fenton, Stoke-on-Trent, Staffordshire
Present location: Albert Island, Royal Docks, Newham, London
1981–2, relocated to present position c. 1989;
Michael Hopkins & Partners
Structural engineer: Anthony Hunt Associates

Patera was conceived by Nigel Dale as a flat-pack, demountable building. Alongside selling kits, he envisaged serviced industrial estates where units could be modified to tenants' requirements. Based in the Staffordshire Potteries, he persuaded his family firm, Longton Industrial Holdings, to set up a subsidiary. Hopkins was approached after Dale discovered that he was developing an aluminium building system with Tony Hunt and Ian Ritchie.[97]

Patera was to be all-steel, with external welded tubular trusses connected by cast pin joints. Pressed-steel sandwich panels incorporating insulation and services served as roof and walls, with glazed ends. The project engineer, Mark Whitby, devised a portal frame with a central connection that acted as a hinge under dead loads and as a rigid joint in wind uplift conditions. Two demonstration units were erected outside the factory by the men who fabricated them. The kit cost £30,000, weighed under 29 tonnes and packed into two 12m (40ft) shipping containers. Like IBM **Cosham** it could be erected on a concrete slab with a fork-lift truck in ten days.

After the parent company withdrew from the venture, Hopkins bought up the rights and introduced modifications. In 1984 two Patera prototypes were re-erected at the West India Docks as exhibition buildings. One survives at the Royal Docks, latterly serving as a boat repair workshop; at the time of writing it is partially dismantled and faces an uncertain future. Another can be found at the **Hopkins Practice Office** in Marylebone.

Further reading:

Nigel Dale, 2024, *The Patera Building: an Enduring Curio*. Madeley: Nigel Dale.

INMOS

Cardiff Road, Duffryn, Newport, Gwent
1981–2; Richard Rogers Partnership
Structural engineer: Anthony Hunt Associates

In 1982 *The Economist* reported on Britain's 'Sunrise Strip' of high-technology industries, which stretched along the newly completed M4 from South Wales to Heathrow Airport.[98] The presence of existing transport infrastructure and research establishments attracted transnational corporations such as Hewlett Packard (Bristol, Reading), Intel (Swindon), Sony (Bridgend) and IBM (Hounslow) as well as homegrown operations. Inmos was founded in 1978 by the computer engineer Iann Barron to manufacture semiconductors for electronic devices. Funding was secured from the National Enterprise Board on condition that the factory was built in the 'assisted area' of South Wales rather than the firm's Bristol base.

Inmos's requirements for sterile production areas, flexible servicing and future expansion suggested a design with outdoors structure and services. A row of nine H-towers defines a circulation spine that separates the production zone from support functions. Air-handling packages are mounted above this internal street, while a series of 38m-span (125ft) trusses are supported by the towers and raking struts. The tubular steel girders were welded off-site and assembled with pinned connections.

Although the combination of logic and drama recalls the nearby Newport Transporter Bridge (1906; Ferdinand Arnodin), Inmos's precursor is the Aram Module (1971), a Rogers/Hunt collaboration for a prefabricated hospital suspended from a masted structure. In turn, Inmos informed the PA Technology Laboratories at Princeton, whose central A-frame was designed with Ove Arup & Partners. Four decades on, the Newport plant is the UK'S largest manufacturer of semiconductors; in March 2024 the government approved its takeover by the US-based Vishay Intertechnology.

09

RENAULT DISTRIBUTION CENTRE

The Spectrum Building, Mead Way, Westmead, Swindon, Wiltshire
1981–3; Foster Associates
Structure and services engineer: Ove Arup & Partners
Listed at grade II* in 2013

Most Foster works are exercises in refinement and fine-tuning from project to project. Renault seems to have been started from scratch and the result is an uninhibited one-off. Sébastien de la Selle, the firm's *co-ordinateur d'expression visuelle*, appointed Foster Associates to design a national parts distribution centre and regional office after seeing their Olsen buildings at Millwall Dock, **Willis Faber & Dumas** and IBM **Greenford**.

The concept of a modular umbrella roof came from Ken Anthony of Ove Arup & Partners who, in exploring design options with Foster's, sketched a single mast from which radiating beams were suspended. This evolved into a two-way series of cable-stayed portal frames, painted the yolky yellow of Renault's diamond badge. The cladding (glass for the offices, metal panels for the warehouse) is set 2m (6.5ft) behind the perimeter structure, whose cranked beams suggest the rhythm of a chassis on an assembly line. The undulating roofscape of flattened and truncated pyramids, into which are let north lights, is clearly legible from inside as well as from the air. An open-plan warehouse accounts for 36 of a total of 42 bays. To the south the bays crumble down to offices, a double-height 'gallery' and the single open bay of a *porte cochère*.

Since Renault moved out in 2001, the building's modular structure has facilitated its flexible subdivision for multiple commercial tenants. A programme of refurbishment was undertaken in 2023–4 at a cost of £4.7 million.

Further reading:

Chris Abel, 1991, *Renault Centre, Swindon*. London: Architecture Design and Technology Press.

HERMAN MILLER DISTRIBUTION CENTRE

Herman Miller Estate, Bath Road, Chippenham, Wiltshire
1982–3; Nicholas Grimshaw & Partners
Structural engineer: Peter Brett Associates

After Herman Miller UK failed to expand their manufacturing site outside Bath, they purchased a 14-acre site at Chippenham, 13 miles distant. They returned to Nicholas Grimshaw & Partners to oversee a phased building programme, starting with a warehouse. An initial scheme, based upon an external lattice structure, was turned down at tender stage by the client's pension-fund backers for its 'unmarketable appearance'.[99]

Project architect Neven Sidor reverted to the familiar concept of a discrete steel frame clad in modular panels that could be swapped out. Rather than the sandwich panels of **Herman Miller Bath**, he opted for separate inner and outer skins, combining bespoke and off-the-peg components. External panels of ribbed aluminium (resembling the bodywork of the Citroën H van) are finished with bright blue powder coating. Instead of a four-way joint, they are fixed onto vertical neoprene gaskets and horizontally lapped and bolted to welded pairs of T-bars. (These mullions were reprised at the ***Financial Times* Print Works**.) On the inside are acoustic liner trays backed with rockwool slabs.

The plan is compact, with loading bays along the long south-west elevation. The adjacent frontage to Bath Road is more talkative, with a curved lavatory pod and office suite enclosed by two curved entrance porches punctuated by bright red piloti. After Herman Miller moved out, the 'blue building' was refurbished (uprating its energy performance certificate from E to A) and painted regulation grey.

SCHLUMBERGER

Schlumberger Gould Research Centre, High Cross,
Madingley Road, Cambridge
1983–5, 1991–2; Michael Hopkins & Partners
Structural engineer: Anthony Hunt Associates; Ove Arup & Partners (roof membrane); Buro Happold (phase II)
Phase I listed at grade II* in 2017

Schlumberger is the supreme monument to the oil age. Founded in 1926 by the brothers Conrad and Marcel Schlumberger of Alsace, the firm became an international conglomerate providing technical services to the petroleum industry. The firm commissioned Philip Johnson to design their Research Center in Ridgefield, Connecticut (1952), while Renzo Piano remodelled their Paris base in 1981–4 (including a covered route with a Teflon roof). Schlumberger's Cambridge facility specialized in drilling mechanics, wellbore physics, fluid mechanics and mathematical modelling. To foster interaction between the 60 researchers, Hopkins put the drilling test station at the heart of the plan, next to a 'winter garden'. Initial plans for a space-frame roof were dropped in favour of a three-bay roof of Teflon-coated glass fibre, engineered by Arup's. The translucent tent combines daylighting with the ability to safely blow out in the event of an explosion. A spider's web of cables suspends the membrane from four masted structures. This sinister circus tent is hemmed in by low office and laboratory wings with triangulated trusses.

By the 1990s, computer modelling was superseding test rigs, so the phase II reception building has a more conventional arrangement of labs and offices. A pair of quadrangles (based on Hopkins' headquarters for **Solid State Logic**) are held apart by a central entrance hall with an air-cushion roof by Happold's. They incorporate coffered floor slabs whose ribbed soffit, like Pier Luigi Nervi's Gatti wool factory in Rome (1951), traces the lines of force through the structure.

Further reading:

David Jenkins, 1993, *Schlumberger Cambridge Research Centre: Michael Hopkins and Partners*. London: Phaidon.

LINN PRODUCTS

Glasgow Road, Waterfoot, Glasgow
1985–7; Richard Rogers Partnership
Structural engineer: Ove Arup & Partners

Linn Products is a high-end audio manufacturer founded in 1973 by the Gorbals-born Ivor Tiefenbrun. He started with the Sondek LP12 turntable, a precision-engineered machine of uncluttered appearance. Tiefenbrun wanted a factory that followed these principles. Designed as the **Lloyd's Building** was being fitted out, RRP's design represents a return to the elegant sheddery of their early work.

Situated in the quiet of the former Eaglesham House estate outside Glasgow, the complex comprises two parts: a low pavilion connected to a narrow, tall warehouse. The main building is sunk into a banked site, with the ground falling away to the east to allow a double-height foyer. The *piano nobile* is a double-height, largely open-plan volume housing production and offices. Research and development, computer suite, plant and an acoustically isolated demonstration room are placed in an undercroft. The structure is a two-way spanning grid on a 14.4m (47ft) module with deep trusses that carry the services.

A link passage gives access to the windowless warehouse, clad in blue corrugated steel. It was designed as a fully automated facility to serve Linn's 'single-stage build' manufacturing process, where skilled operators assemble complete products from start to finish before signing them as a quality assurance measure. Three aisles are patrolled by automated guided vehicles that retrieve components and collect the finished items. The robotic carts (Huey, Dewey and Louie) were still in active service in 2024. The facility was extended with north and south wings in 2002–3.

LADKARN HAULAGE

Solar House, Alpine Way, London Industrial Park, Beckton, London
Original location: West India Docks, Isle of Dogs, London
1985, relocated 1987–8; Nicholas Grimshaw & Partners
Structural and services engineer: Ove Arup & Partners
Structural engineer for re-erection: Trigram Design Partnership

Demountability is a central principle of high-tech architecture – but it is seldom put to the test. Along with the **Patera prototype**, Ladkarn is one of few buildings to change its address. It took almost two years of negotiation with the London Docklands Development Corporation (LDDC) before the firm signed a 25-year lease on a site in West India Docks. As a specialist in plant hire and groundworks subcontracting, Ladkarn's brief was straightforward: head offices and a clear-span vehicle maintenance workshop. From a list of approved architects supplied by the LDDC they chose Grimshaw's, who for a fixed fee devised a masted structure 'rather like the bow sprits of old sailing ships'.[100] Its scarlet outrigging lies in a sequence of Grimshaw exoskeletons that starts with the **IBM Sports Hall** and the **Oxford Ice Rink**.

As the building neared completion it became clear that it would have to make way for Canary Wharf, approved by the LDDC in October 1985. After accepting an offer they couldn't refuse, Ladkarn unbolted the structure and re-erected it three miles east at Beckton. This was made easier by the fact that the client also acted as contractor. Most of the structural frame, the aluminium windows, partitions and the boiler could be reclaimed, but the 'crinkly tin' cladding proved too warped and dented to be salvaged. Some modifications were necessary outside the low-regulation Enterprise Zone – the pin-jointed internal columns would not take an intumescent coating, so they had to be replaced – in all, the reuse saved about a third of the cost of putting up the same design from scratch.

Further reading:

Alan Cox, 1995, *Docklands in the Making: The Redevelopment of the Isle of Dogs, 1981–1995*. London: Athlone Press.

SOLAR
HOUSE
SOLAR HOUSE

FINANCIAL TIMES PRINT WORKS

East India Dock House, 240 East India Dock Road, Poplar, London
1987–8; Nicholas Grimshaw & Partners
Structural engineer: J Robinson & Sons
Press engineer/interior consultants: Robinson Design Partnership
Listed at grade II* in 2016

The *Financial Times* was one of the last dailies to abandon their City premises (**Bracken House**) to build a new print works in the Isle of Dogs Enterprise Zone. The newspaper had long championed higher standards of industrial architecture through an annual award scheme but here was an opportunity for direct patronage. Grimshaw's practice was probably appointed on the recommendation of the FT's architecture correspondent, Colin Amery. Their layout allowed an effortless flow of pink newsprint from delivery and storage at the west end, along a north-facing press hall to the despatch of bales of papers at the other end. By night, fully glazed walls showed off the rolling conveyors of the web-offset presses to passing motorists in the tradition of Owen Williams's *Daily Express* building in Manchester.

The point-fixed, tinted glazing, based on Pilkington's Planar system, marked a new level of sophistication in Grimshaw's industrial buildings. Four 2 × 2m (21 × 21ft) panes are bolted to stainless-steel 'dinner plates', which connect to tension rods. Wind loads are resisted by aerofoil columns attached to glazing arms. The blind outer bays are clad in ribbed panels of superplastic aluminium, with smooth and glazed variants for the external stair pods.

Remarkably, the presses rolled for only seven years before printing was outsourced to the West Ferry plant, a joint venture of the *Telegraph* and *Express* groups. Afterwards the print works was converted to its digital equivalent: a data centre. East India Dock House is today a Poplar landmark, joining Goldfinger's Balfron tower and outlasting the Smithsons' Robin Hood Gardens.

Further reading:

David Jenkins, 1991, *Financial Times Print Works*. London: Architecture Design and Technology Press.

T INDIA DOCK HOUSE

REUTERS DATA CENTRE

Telehouse South, 1 Blackwall Way, Paul Julius Close, Poplar, London
1987–9; Richard Rogers Partnership
Structural engineer: Ove Arup & Partners
Services engineer: Jaros Baum & Bolles

In 1984 the London Docklands Development Corporation commissioned RRP to prepare a masterplan for the former Royal Docks, at 400 hectares the largest of London's dock complexes and the last to be redeveloped. Following this the practice worked up a retail and leisure proposal at the Royal Albert Docks for the developer Stuart Lipton. Neither materialized (although Rogers later designed one of a series of stormwater pumping stations for the LDDC). Lipton came back with another waterside site: Blackwall Yard, a shipyard of 17th-century origin that ceased operating in 1987. Only phase one was realized, a secure data centre for the Reuters news agency, its interior fitted out by Fitzroy Robinson.

This reworks the **Lloyd's** diagram of deep, open floors sprouting perimeter lifts, service risers and curved glass staircases. The elevations are layered into strata, with a rooftop cage of satellite dishes above offices above plant rooms. Plans can be reconfigured from office to data storage uses, with interchangeable solid and glazed panels. Early data centres were supervised by a relatively large support staff, hence the carefully composed entrance, restaurant and recreation annex (demolished in 2020), hard landscaping and the refurbishment of the late 18th-century eastern dry dock. In 2016 a lower annex was added to the facility, which today belongs to a network of collocated data centres in Docklands. The remainder of Blackwell Yard, long used as a car park, is now being regenerated with a mixed-use development.

MCLAREN TECHNOLOGY CENTRE

Chertsey Road, Woking, Surrey
1998–2003; Foster + Partners
Structural engineer: Arup

While galleries and cultural institutions acquire derelict artfully distressed former industrial buildings, 21st-century factories become white boxes. The production facilities for Rolls-Royce at Goodwood (2001–3; Nicholas Grimshaw & Partners) or VW's Gläserne Manufaktur in Dresden (2001; Henn Architekten) are about branding, display and the visitor experience.

So it is with McClaren, whose headquarters consolidate about 1,500 staff formerly spread across 14 locations around Woking. It is a low-rise disc sunk into the landscape, its yin-yang plan completed by an artificial lake that helps to cool a 145m-long (475ft) wind tunnel. The sinuous waterside façade has a retro-futuristic quality, like a slinky version of the De La Warr Pavilion. Aluminium 'wind blades', designed in collaboration with McLaren's engineers, transfer wind loads from the full-height glass wall to inset columns. The production floor is organized around 18m-wide (60ft) bays separated by internal streets that stretch towards a full-height lakeside boulevard. Above are top-lit offices and design studios and a raised walkway that forms part of a VIP route.

In 2010–11 the Technology Centre was joined by the playing-card-shaped Production Centre, designed for the assembly of McLaren 12C sports cars. In 2022 McLaren sold the complex to free up equity, staying on as a tenant on a 20-year lease.

ERCOL FACTORY

Summerleys Road, Princes Risborough, Buckinghamshire
2001–2; Horden Cherry Lee Architects
Structural engineer: WSP South

What should a factory look like in a post-industrial economy? Deluxe-handicraft, like David Mellor's Hathersage cutlery works (1988–9), or the white-floored labs of the **McLaren Technology Centre**? Never radical, the Ercol range of furniture sought to reconcile mass production with homegrown materials and craftsmanship, and this middle way seems to have permeated the design of its factory. Ercol was established in 1920 as Furniture Industries, later renamed after its founder Lucian R Ercolani (1888–1976). At its post-war peak, the High Wycombe factory employed some 800 workers and produced 2,000 pieces of furniture daily, using elm and beech from the company's Hertfordshire woodlands.

Still family-run, the firm resolved to keep its manufacturing base in the UK, and in 2002 moved its 350 employees to a new factory. Its site, on the edge of the Chilterns, was previously occupied by the Building Research Establishment's timber laboratory. The phasing out of solvent-based finishes and a sophisticated dust-extraction system allowed a single production space. An open-plan, double-height manufacturing hall incorporates a top-lit circulation spine, with a 120m (395ft) strip window providing woodland views. A white steel colonnade throws the profiled metal shed into shade, recalling the elegant trabeation of Raphael Soriano and Roche Dinkeloo. It breaks forward to order an intricate sequence of front-of-house spaces, including entrance lobby, showroom, double-height gallery and a restaurant beyond. This framing device, poetically described by Horden as 'moving through layers in a forest', was influenced by Renzo Piano's Menil Collection in Houston, Texas, although a closer precedent is Horden's **Yacht House**.[101]

High Street
Taxis
Trains
Bus stops
Information
Shopping Centres
IN CINEMAS
APRIL 18
STREAM GOATLOADS OF
BLOCKBUSTER ENTERTAINMENT

8 | transport

en Elizabeth Olympic Park
Westfield Stratford City
E RAISING PRICES AGAIN.
← High Street
Taxis →
Information →

BEDFORD MIDLAND STATION

Ashburnham Road, Bedford, Bedfordshire
1976–8; British Rail London Midland Region architect's department

'Bright, airy and logical, almost High Tech,' was Jonathan Glancey's verdict on Bedford's £1 million station.[102] Its bronzed, gasket-glazed curtain walling and Triodetic aluminium space-frame roof shows the influence of IBM **Cosham** and **Willis Faber & Dumas** on BR's project architect John Sanders. New track alignments, part of the electrification of the London to Bedford line, required the rebuilding of the station on a new site. As the tracks came in at a higher level it was necessary to raise the glass box on a podium, designed with landscape architects the Michael Brown Partnership as a hard-wearing base of red-brick steps, ramps and low walls.

At the time of its opening by BR chairman Peter Parker, amenities included a ticket office, travel centre and a 'travellers-fare buffet'. Little is now left of that interior, but the waiting room retains its Robin Day Polo chairs and the concourse is still crowned by its jaunty space frame. Bedford has a smaller sibling at Radlett, built in 1978–80, and a space frame was used in BR's rebuilding of London Bridge Station (1976–8, demolished between 2009 and 2017). It is likely that Bedford Station will again be rebuilt as part of East West Rail, the planned new line between Oxford and Cambridge.

Pages 242–3: Stratford Bus Station, Newham, London, 1994, London Transport.

STANSTED TERMINAL

Stansted Airport, Essex

1986–91; Foster Associates

Structural engineer, fire engineering, drainage: Ove Arup & Partners

Stansted turned airport design on its head. Up until then, terminals were warrens of levels and zones, with services stuffed above ceiling tiles. At London's long-awaited third airport, the 'servant' functions – plant, baggage handling and railway interchange – are displaced into a vast undercroft on the same principle as Richard Rogers' **PA Technology**. Above is a single open volume, with daylight filtering through a lightweight canopy. Departures and arrivals are side by side, providing a linear progression between landside and airside. Glimpses of take-off are snatched through fully glazed walls.

In response to early uncertainty about permitted capacity, Foster Associates laid out a $36m^2$ ($388ft^2$) grid capable of phased expansion. Structural 'trees' support an origami roofscape that clearly descends from the **Renault Distribution Centre**. Air, light and power are routed from the undercroft through the freestanding trunks. Their service pods, bristling with ducts, screens and CCTV, are pieces of multi-functional hardware. This leaves the roof with little more to do than keep the weather out, draw in daylight and at night, bounce back light from uplighters in the trees.

Stansted has grown gracefully, its capacity increasing from 8 million passengers per year to a peak of 28 million in 2019. Today, the rites of security and consumption are all-pervasive; for departing passengers, the sense of a direct, legible route has become a gastro-intestinal tract of duty-free, mall and food court. Yet the architectural values of lucidity, openness and calm efficiency just about win out under that luminous roof.

WATERLOO INTERNATIONAL TERMINAL

Platforms 20–24, Waterloo Station, York Road, London
1990–3; Nicholas Grimshaw & Partners
Engineers: British Railways Board Network Civil Engineers; YRM Anthony Hunt Associates (roof); Cass Hayward and Partners (track support); Sir Alexander Gibb and Partners (basement, arches)

The Channel Tunnel Rail Link, enacted by Parliament in 1987, started with a typically British fudge. As there was no immediate prospect of a preferred route through Kent to north London, an additional terminus was grafted onto Waterloo Station. With the construction of the 'Stewarts Lane Chord', the Eurostar service shared the Kent-bound lines from Victoria, only reaching top speeds of 300km/h (186mph) on the other side of the Channel. More inspired was the decision taken by Jane Priestman, British Rail's director of architecture, design and environment, to appoint Nicholas Grimshaw & Partners (over Foster Associates and Ahrends Burton and Koralek) to design Waterloo International. Promoted as a 'gateway to Europe', it was BR's first new terminus since Euston and the organization's last *grand projet* before privatization.

Eurostar was in direct competition with air travel: the terminal was designed to handle 15 million passengers a year, the same number as **Stansted** with its planned expansions. Its trains had a capacity of 750, almost double that of a passenger jet. Grimshaw's challenge was how to move people to and from the trains and through security, immigration and customs as efficiently and graciously as possible. Arrivals and departures were set on different levels, forming a two-storey viaduct that supported the platforms. A basement car park acted as a foundation raft that spanned the shallow Underground lines below. Above sailed an asymmetrical roof structure of glass and welded steel. This charismatic shed was the public face of Eurostar – yet it was only transiently experienced by its passengers, who were penned in the departures lounge until the last minute.

In order that it could swallow whole the 18-carriage trains, the roof was 400m (1,300ft) long. Curved and tapered like a giant windsock, it had 48m (157ft) spans towards Waterloo Road, diminishing to 32m (105ft) at the Paris end. Distant descendants of the blue-painted exoskeleton of **IBM Hursley**, the trusses are flattened bow-string arches, each with three pinned connections. The middle

pin is shifted over the centreline, allowing the roof to rise steeply to clear a single track hard against the western edge. On this side, a glass roof is suspended from an external truss. On the other, the trusses are set under corrugated stainless steel cladding. Instead of specially cut panes of glass, standard sheets overlap top and bottom like armadillo scales. Concertina-shaped neoprene gaskets, set between each pane, flex to accommodate varying alignments and absorb the vibration of braking trains. The curvilinear geometry is resolved with just three stainless-steel castings that can be joined at the desired angle with cogged connections. Unusually, the practice designed all the fitting and fixtures, including the ticket booths, public telephones, toilets and lighting fixtures. Signage employing Jock Kinneir and Margaret Calvert's Rail Alphabet and the Design Research Unit's pictograms established continuity with British Rail modernism.

Handed over in May 1993, over a year ahead of the opening of the terminus, Waterloo International was hailed as 'Britain's last great railway station of the twentieth century and a confident landmark in the great tradition of railway engineering'.[103] Along with the British Pavilion in Expo '92 in Seville, it marked the maturation of Grimshaw's design approach and his emergence as a global player. Yet only two years after Eurostar opened, Parliament approved the construction of a high-speed line from St Pancras. When the final whistle blew at Waterloo International on 13 November 2007, it had served almost 82 million passengers over 13 years (cost per passenger journey: £1.58). It took over a decade before platforms 20 to 24 were permanently brought back into domestic service, while in 2022–3 the terminal substructure was converted to a shopping mall and 'experience-led destination' named The Sidings.

COSTA
BURGER KING

EAST CROYDON STATION

Station Approach, Croydon
1990–2; Alan Brookes Associates
Structural engineer (superstructure): YRM Anthony Hunt Associates

While much of central Croydon was rebuilt with office blocks in the 1960s, the town's commuters were left with the station erected in 1894–5 by the London, Brighton and South Coast Railway. By the 1980s it was clear that Croydon's principal interchange was struggling to cope with 15 million passengers a year. The stumbling block to rebuilding was always that it would cause a massive service disruption across the network. This alone stymied the ambitions of British Rail's property board to realize an 'air rights' development above the platforms. Instead, BR's in-house architects, led by Nick Derbyshire with Tony Hunt, devised a scheme in which a long-span masted structure could be slung across six railway tracks in one go. This was taken forward by Alan Brookes Associates, a practice that combined architectural design with technical consultancy and component development.

Their constructional expertise tells in the legible, refined detailing of the booking hall, which is hung beneath the trusses. It is fully glazed with big sheets of toughened glass, each fixed at four points with cast stainless-steel brackets. These connect to elliptical aluminium mullions that provide wind bracing. Differential movement between structure and cladding is taken up by a scissor-arm head assembly. All is lovingly detailed; rather less elegant is the white tubular steel superstructure. Four ladder towers are connected by tie bars to two primary box trusses, with lattice trusses running between. Lightweight ramps of steel and glass descend to the platforms, evoking ship's gangways. Network Rail has proposed rebuilding the station 100m (330ft) to the north to ease congestion and support Croydon's regeneration.

rac
rac
rac

RAC SUPERCENTRE

Great Park Road, Bradley Stoke, Bristol
1993–4; Nicholas Grimshaw & Partners
Structural engineer: Alan Baxter Associates

In response to increasing competition, the RAC decided to modernize its operations in the mid-1980s, investing in a computer-aided rescue service (CARS) and building five regional control centres. These 'supercentres' were to be statement buildings, conspicuously located at motorway locations to symbolize the RAC's rebranding as a high-tech service organization. An earlier example at Walsall, off the M6 (1989–90, Building Design Partnership) provided a template for a technologically sophisticated nerve centre. It was followed by Grimshaw's supercentre, designed to be taken in at 70mph (112km/h) from the Almondsbury Interchange at the junction of the M4 with the M5. The building is a rounded triangle from which rises a 65m-high (215ft) mast with an observation gallery – a miniature version of Foster's Torre de Collserola in Barcelona. It was explained to RAC staff as a Dairylea triangle topped with an olive on a cocktail stick.

While certain concepts are refined from **Western Morning News** – particularly the tilted and tinted glazing, control tower and central atrium – the result is sleekly futuristic. Shelter from motorway noise and pollution was sought by sinking the building into a landscaped bowl, with a footbridge leading to the main entrance on a *piano nobile*. The full-height atrium was conceived as a social 'mixing valve', with a dramatic sequence of flying staircases encouraging chance conversations. External sunshades, suspended from stainless-steel cables, double as walkways for the cleaning and maintenance of the glazing. Electric roller blinds were later installed to counter glare and solar gain.

WESTMINSTER STATION

Bridge Street, Westminster, London
1993–9; Michael Hopkins & Partners
Civil and structural engineer: G Maunsell & Partners

Westminster was the most sensitive and complex project on Roland Paoletti's Jubilee Line Extension. After a new parliamentary building was approved in 1990 Michael Hopkins & Partners was the obvious choice to integrate the new station with his **Portcullis House** above. (An initial scheme had two ticket halls, one for the existing Circle and District Line and another beneath Parliament Square.) To reduce construction impacts, the ticket hall and the main circulation were placed in a deep-level, diaphragm-walled box. It was a technique pioneered at the Hong Kong Mass Transit Railway on which Paoletti was chief architect. It was decided to open up the box, replacing internal compartments with central columns and 'flying shores', colossal horizontal braces of solid steel (Ian Ritchie's Bermondsey Station repeats the trick with concrete horizontal trusses). Outside the box, the JLE tunnels were stacked eastbound above westbound to skirt Big Ben.

Between the tight spaces of the ticket hall and platforms lies a vast underground atrium with 17 escalators and five lifts threaded through the openwork structure. Lighting highlights a sense of subdued drama, with washes and spotlights picking up the mica sparkle of fair-faced concrete and the rude castings of the box. Poles apart from the isotropic space of the serviced shed, this is a progression through composed space (albeit one composed from civil engineering), a descent into a Piranesian underworld. Westminster embodies Paoletti's vision of the JLE as 'an engineering that expresses itself as architecture; an engineering in which people can delight'.[104]

CANARY WHARF STATION

Bank Street, Canary Wharf, London
1994–8; Foster Associates
Civil engineer: Posford Duvivier, De Leuw Chadwick;
structural engineer: Ove Arup & Partners

Serving 40 million passengers a year, this station is the showpiece of the Jubilee Line Extension (JLE). Which is fitting, as it was the redevelopment proposals for Canary Wharf that swayed the Thatcher government into green-lighting the long-postponed tube project. The station box occupies the infilled middle dock of the old West India Docks, its 24m (80ft) depth determined by the level of the adjacent Thames tunnels and its length of 280m (920ft) by the need for a 'scissor crossover' permitting trains to switch track and reverse. On top is a public park, nibbled east and west by station entrances. Their glazed canopies shelter the station from weather while throwing light deep underground. They resemble miniature halves of the **National Botanic Garden of Wales**, while a subsidiary eastern entrance apes the *Fosterito* canopies of the practice's Bilbao Metro.

Banks of escalators descend to an open-plan concourse, lined with passenger amenities, before dropping down to the trains. The route is direct, legible, orderly – yet awe-inspiring. A giant order of elliptical concrete columns rises from platform to roof, capped with bearings that permit structural movement. Combustible or fragile materials were ruled out, so fair-faced concrete dominates, particularly in the gull-wing vaulted roof. At platform level the concrete diaphragm walls of the box are left exposed. The JLE's designers enjoyed something their predecessors lacked: the luxury of space. Space for circulation, safety requirements, future expansion, but above all its architectural possibilities. At Canary Wharf the passenger moves calmly through one of London's great public interiors.

Way out
Jubilee Place
DLR
Elizabeth

STRATFORD MARKET DEPOT

Burford Road, Stratford, London
1994–6; Chris Wilkinson Architects
Structural engineer: Acer Consultants

The year 1991 was an *annus mirabilis* for Chris Wilkinson. Not only was his *Supersheds* published but he got the chance to design one – his first major commission. A rolling stock depot was needed at the end of the Jubilee Line Extension to supplement the one at Neasden. Its site had previously been occupied by the Cistercian abbey of Stratford Langthorne and, from 1879 to 1991, an iron-framed wholesale market established by the Great Eastern Railway. An outline design was completed over two weeks of intensive sessions with project engineer Richard Fenton.

It was decided to locate the main structures to the north to avoid the archaeological remains of the precinct, over which would run sidings. The track alignment suggested a 30-degree parallelogram form and this, together with a requirement for 8m (26ft) headroom, led to a diagrid roof structure on a lazy arc. The curve was too shallow for a clear span so a triple-arch sequence was adopted, bearing upon two rows of 'tree' columns. Tucked under the eaves are exposed V-shaped columns, hinged top and bottom to allow thermal expansion. Roof members are picked out in pale green in tribute to the Runcorn–Widnes Bridge over the Mersey. The 100 × 190m (330 × 625ft) volume is evenly lit by rooflight strips, lateral clerestories, a south wall of translucent GRP and structural glazing to the north.

Although inaccessible to passengers (it can be glimpsed from the Docklands Light Railway between Stratford High Street and Abbey Road), Wilkinson's shed is nonetheless a high point of Roland Paoletti's enlightened patronage of the JLE.

Jubilee line
96003

NORTH GREENWICH STATION

5 Millennium Way, Greenwich, London
JLE station: 1995–9; Alsop Lyall & Störmer with JLE Project Architects
Civil and structural engineer: Benaim Works JV
Bus station: 1997–8, Foster Associates
Structural engineer: Anthony Hunt Associates

North Greenwich was an afterthought on the Jubilee Line Extension: it didn't feature in the 1989 parliamentary bill and a reroute was only agreed after an offer of financial support from the developers of the Greenwich Peninsula. A few years later, the former gasworks site was chosen as the centre of the nation's millennium celebrations.

Constructed within a concrete box 400m (1,300ft) long, 30m (100ft) wide and 25m (80ft) deep, the station is the largest on the JLE. Instead of an intermediate floorplate, Alsop Lyall & Störmer placed the concourse in a parabolic 'boat' suspended from the roof on struts. (An early scheme went one step further and omitted the roof, but this was vetoed as prejudicial to a future 'air rights' development.) A service spine and smoke extract duct are stowed beneath. Two rows of raking 'v' columns, elliptical in section, hold the roof up and help the base resist water pressure. Walkways and escalators negotiate vertiginous slots of space. Alsop wanted colour and lighting to create a feeling of endless subterranean space. Ultramarine mosaic envelops the columns, while backlit glass walls of cobalt blue unify the two levels.

The Underground station is accessed from the spine of a bus interchange by Foster Associates. Resembling a horseshoe crab on plan, it is dominated by a swooping roof with a 160m (525ft) span, allowing natural ventilation and unobstructed views. Foster's interchange is set to be demolished as part of a new masterplan for Greenwich Peninsula approved in 2020.

NORTH GREENWICH

CANADA WATER STATION

Deal Porters' Way, Southwark, London
Underground station: 1995–9; JLE Project Architects with Herron Associates
Bus station: 1997–9; Eva Jiřičná Architects
Civil and structural engineer: Benaim Works JV; structural engineer for rotunda: Buro Happold

Standing at the edge of the infilled Surrey Commercial Dock, Canada Water Station brought new transport links to an isolated part of Docklands. The interchange was a pilot study for the Jubilee Line Extension and follows the stripped-down specification of the Hong Kong Mass Transit Railway. The JLE tracks are 19m (62ft) below street level, sufficiently shallow to make cut-and-cover construction viable but deep enough to fit under the existing East London line. A steel and glass drum, on the model of Holden's Arnos Grove or the entrance shaft to Brunel's nearby Thames Tunnel, acts as a light scoop and a beacon by night (page 4). As at Southwark and Bermondsey, the station box was designed to allow a multi-storey 'air rights' structure to be constructed over it. Finishes are austere and industrial: in-situ concrete, grey mosaic and black mesh trays for overhead services.

If the JLE station represents engineering unadorned, Eva Jiřičná's bus station takes a simple diagram and adds layers of detail. It is a linear shelter, with perimeter stops and a stone-clad core accommodating a control centre and passenger facilities. As the station is overlooked by the towers of the Canada Estate (1962–4, London County Council) it was necessary to screen residents from engine noise and emissions as well as protecting passengers from rain and London pigeons. A pair of aerofoil wings, likened by Jiřičná to a bird in flight, are cantilevered from a glazed spine, rhomboid in profile and bearing upon five pin-jointed columns. Towards the tube station the wing cleaves away, leaving a low glass canopy to skirt the drum.

SUB
SIDE
DRINK
EH219
CLEANERAIR

STRATFORD STATION

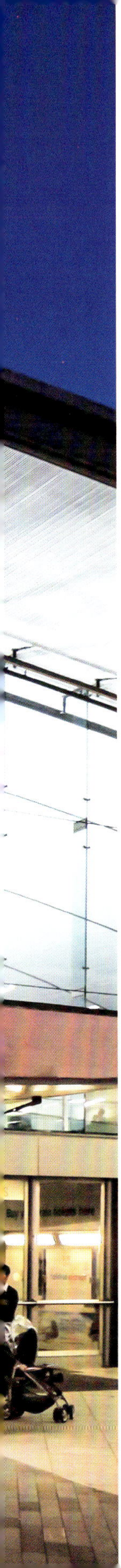

STRATFORD STATION

Station Street, Stratford, Newham, London
Concourse: 1996–8; Chris Wilkinson Architects
Structural and services engineer: Acer Consultants;
Civil engineer: Ove Arup & Partners
JLE platforms and staff building: Troughton McAslan
Structural and civil engineer: WSP Group
Bus station: 1994; London Transport
Structural and civil engineer: Anthony Hunt Associates

Stratford is the end of the line for the Jubilee Line Extension. The initial proposal was limited to a pair of JLE platforms and a staff building (realized in modified form by Troughton McAslan). But proposals for Stratford's regeneration, including the possibility of Crossrail and Channel Tunnel Rail Link services, justified a more ambitious approach. In 1994, with **Stratford Market Depot** on site, Chris Wilkinson Architects were brought on board to design the front end of a multi-modal transport interchange.

A multi-level concourse is housed within a grand volume, connected to all lines but addressing the town. An asymmetric roof, resembling a quarter ellipse, is cantilevered on tapering ribs from the trackside. It rises to a slanted structural glass wall, lifted clear of the ground upon four piers. Despite echoes of 19th-century train sheds and the Galerie des Machines in Paris, Stratford is redolent of an airport terminal – Washington Dulles, say – where one experiences a disconnect between uplifting arrival and a disorienting boarding process. Disconcertingly, the North London Line barges through the concourse, bridged by an upper walkway. Major improvements preceded the 2012 Olympics, including new lifts and platforms, a northern entrance and a pedestrian bridge to Stratford International Station and the Westfield shopping centre. Further reforms are planned.

Notwithstanding its ephemeral appearance, the tensile fabric roof of the bus station outside (pages 242–243) predates Wilkinson's station. Designed in-house by Soji Abass of London Transport and engineered by Tony Hunt, it is a gossamer row of inverted canopies, resembling Gothic fan vaulting in Teflon.

HEATHROW T5

Wallis Road, Longford, Hounslow
2002–8; Richard Rogers Partnership
Structural engineer: Ove Arup & Partners

Terminal 5 was won at competition in 1989 but after an interminable passage through the planning system, the first take-off was delayed until March 2008. In the meantime, increased security and retail requirements and a reduced land-take prompted the evolution of RRP'S scheme. The competition entry, strongly influenced by Foster's Stansted, had departing and arriving passengers on the same level, sharing that staple of 1990s architecture, the wavy roof. Transit facilities and baggage handling were placed in an undercroft. Site constraints prompted a deeper section and the vertical segregation of arrivals and departures. RRP proposed a series of 'canyons' to bring daylight to the lower storeys but these were dropped after concerns about flexibility and security (the concept survived at Rogers' Barajas T4 in Madrid).

The approved scheme for Heathrow reverts to a loose-fit strategy characteristic of Rogers. The floorplates are structurally independent of and set back from the roof structure, a series of single-span, double-curved trusses that spring from inclined 'tree' columns. As airport regulations precluded the use of conventional cranes, whole sections of roof were assembled on the apron and jacked into position. All departing passengers, whether travelling by train, tube or car, come via a landscaped 'interchange plaza', between the terminal and a six-storey building that contains a bus station and multi-storey car park. The check-in hall and departure lounge are placed underneath the trusses, while arriving passengers pass along a full-height perimeter route.

notes

1 Herbert Read, 1945, 'Machine Aesthetic', in *A Coat of Many Colours*. London: George Routledge & Sons, pp. 320–4.

2 Davies, 1988, p. 6.

3 A W N Pugin, 1853, *The True Principles of Pointed or Christian Architecture*. London: Henry G Bohn, p. 1.

4 Cedric Price, 1979, 'Technology is the Answer But What Was the Question?' A recorded talk for Pidgeon Audiovisual, https://www.pidgeondigital.com/talks/technology-is-the-answer-but-what-was-the-question-/ (accessed 11 Feb. 2025).

5 *Plan*, no. 8, 1950, p. 2.

6 Banham, 1960, p. 10.

7 Oxford English Dictionary, s.v. 'high-tech (n.), sense 1,' https://doi.org/10.1093/OED/3769783460 (accessed 1 June 2024).

8 Robert Maxwell, 1972, *New British Architecture*. London: Thames & Hudson, p. 29.

9 *Progressive Architecture*, July 1978.

10 *Architects' Journal*, 19 September 1979, vol. 170, no. 38, p. 585.

11 Joan Kron and Suzanne Slesin, 1978, *High Tech: the Industrial Style and Source Book for the Home*. New York: C N Potter, p. 1.

12 Kenneth Frampton, 1980, *Modern Architecture: A Critical History*. London: Thames & Hudson, pp. 300–5; Charles Jencks, 1980, *Late-Modern Architecture*, New York: Rizzoli International Publications.

13 *Financial Times*, 30 November 1967, p. 14.

14 Gannon, 2017; Alan Colquhoun, 'Literal and Symbolic Aspects of Technology', *Architectural Design*, vol. 32, November 1962, pp. 508–9.

15 William Mclean, 2018, *Technological Innovation in Architecture; the Role of the Aberrant Practitioner*. PhD thesis, University of Westminster.

16 Fuller's question was famously directed to Norman Foster on a 1978 visit to the Sainsbury Centre and later became the title of a 2010 film directed by Norberto López-Amado and Carlos Carcas. By then it had become something of a catchphrase of Fuller's, as attested by earlier citations (e.g. *The Canadian Architect*, October 1967, vol. 12, no. 10, p. 8). Moreover, Fuller was not the first to pose the question. It formed a 1920s advertising campaign by the Truscon steel company of Youngstown, Ohio ('How much does *your* building weigh? Unnecessary weight squanders dollars. Every extra pound in your building must be paid for three times – in first cost, freight charges and labor of erection', in e.g. *Engineering News Record*, 30 June 1921, p. 7).

17 The first geodesic dome was erected not by Fuller but by the engineer Walther Bauersfeld around 1922 as formwork for the Zeiss Planetarium in Jena, Germany.

18 R Buckminster Fuller, 1969 (reprinted 2008), *Operating Manual for Spaceship Earth*. Baden: Lars Müller Publishers.

19 Gilbert Herbert, 1984, *The Dream of the Factory-Made House: Walter Gropius and Konrad Wachsmann*. Cambridge, Massachusetts: The MIT Press.

20 Peter Smithson, 'Just a Few Chairs and a House: An Essay on the Eames-Aesthetic', *Architectural Design*, September 1966, pp. 444–5; Dirk van den Heuvel and Max Risselada (eds), 2004, *Alison and Peter Smithson – From the House of the Future to a House of Today*. Rotterdam: 010 Publishers, p. 9.

21 This quotation appears on a typographic collage on the back page of the third issue of *Polygon* (1957), the student magazine edited by Outram. See also John Outram and Wilfred Marden, 'The Technology of Change', *Architects' Journal*, vol. 129, no. 3342, 19 March 1969, 455.

22 Reyner Banham, 1976, *Megastructures: Urban Futures of the Recent Past*. London: Thames & Hudson.

23 Stanley Matthews, 2007, *From Agit-Prop to Free Space: The Architecture of Cedric Price*. London: Black Dog Publishing, p. 66.

24 Peter Cook, 'Richard Rogers + Architects', in Barbie Campbell Cole and Ruth Elias Rogers (eds), 1985, *Richard Rogers + Architects*. London: Academy Editions, pp. 4–5 (p. 5).

25 Charlotte Ellis, 'Up and coming in England', *Architectural Review*, vol. 185, no. 1107, May 1989, pp. 26–62.

26 The gerberette was named after Heinrich Gerber (1832–1912), the German civil engineer and inventor of the cantilever girder.

27 Happold recalled that the idea of cast structural joints came from a discussion with Koji Kameya, the partner of Kenzō Tange (Edmund Happold, 'Essential Engineer', *RSA Journal*, vol. 143, no. 5456, January/February 1995, pp. 85–86 (p. 86)).

28 Deyan Sudjic, 'The Man Who Took High Tech Out to Play', *Sunday Times*, 16 January 1983, pp. 26–31.

29 Martin Pawley, 'Good Old Frank', *Architects' Journal*, vol. 183, no. 4, 22 January 1986, pp. 30–31 (p. 30).

30 Frank Newby, 'Hi-tech or Mys-tech', *Transactions of the Royal Institute of British Architects*, vol. 3, no. 6, pp. 18–27; Martin Pawley, 'The Secret Life of the Engineers', *Blueprint*, no. 55, March 1989, pp. 34–36 (p. 36).

31 Peter Rice, 1992, 'RIBA Royal Gold Medal Speech 1992', *Arup Journal*, vol. 27, no. 4, pp. 20–23.

32 Ian Liddell, cited in *Patterns* [the in-house journal of Buro Happold], no. 12, July 1997, p. 6.

33 Mark Whitby, 'Anthony Hunt, 1932–2022', *The Structural Engineer*, vol. 100, no. 10, October 2022, p. 8.

34 Murray Fraser and Joe Kerr, 2007, *Architecture and the 'Special Relationship': The American Influence on Post-War British Architecture*. Abingdon: Routledge.

35 'One plus one', *LA Architect*, February 1975, pp. 2–3 (p. 3).

36 *A New Wave of Japanese Architecture*, Catalogue 10, The Institute for Architecture and Urban Studies, September 1978, New York.

37 '25th Anniversary Tubeworkers Limited', *Tubular Structures*, vol. 36, July 1984, pp. 9–12, with thanks to Jo Smith.

38 'The flat pack thoughts of Chairman Manning', *New Steel Construction*, 1 October 2006, https://www.newsteelconstruction.com/wp/the-flat-pack-thoughts-of-chairman-manning/ (accessed 16 Feb. 2025).

39 Lynda Relph-Knight, 1983, 'Designing in Detail', *Building Design*, no. 665, 11 Nov. 1983, pp. 30–32 (p. 32).

40 Better known is the use of gasket glazing in the General Motors Technical Center, Warren, Michigan (1949–55, Eero Saarinen). Its first large-scale architectural use in England was the Cummins Engine Factory, Darlington, of 1964–5 by the Connecticut-based Kevin Roche, John Dinkeloo and Associates (the successor practice to Eero Saarinen). Neoprene gaskets were taken up by Team 4 at Creekvean, Cornwall (1964–7, grade II*), and widely used by Norman Foster, Richard Rogers and others thereafter.

41 'Secret Life', p. 36.

42 Martin Pawley, 1990, *Theory and Design in the Second Machine Age*. Oxford: Basil Blackwell, p. 40.

43 Andrew Saint, 1987, *Towards a Social Architecture: The Role of School Buildings in Post-war England*. New Haven: Yale University Press.

44 Claude Lévi-Strauss, 1962(reprinted 2021), *Wild Thought: A New Translation of "La Pensée Sauvage"*. Chicago: University of Chicago Press, p. 21.

45 Charlotte Ellis, 'Jean Prouvé: Man of Parts', *Architects' Journal*, 20 Feb. 1985, vol. 181, no 8, pp. 28–31.

46 Brian Waters, 'A Year at Lloyds', *Building*, vol. 247, no. 7361, 21 September 1984, pp. 30–37 (p. 36).

47 Peter Cook, op. cit., p. 5.

48 Michael Hopkins, 'Technology Comes to Town', *RSA Journal*, vol. 140, no. 5429, May 1992, pp. 395–404.

49 Illustrated on the front cover of the *Architectural Review*, vol. 177, no. 1059, May 1985.

50 https://www.dezeen.com/2018/04/25/owen-hatherley-opinion-high-tech-architecture/ (accessed 23 March 2025).

51 A set of minimum space standards recommended by the Parker Morris committee's 1961 report *Homes for Today and Tomorrow*.

52 Humphrey Spender, 1983, *Worktown People: Photographs from Northern England, 1937–38*. Bristol: Falling Wall Press.

53 The *Guardian*, 25 February 2013, https://www.theguardian.com/artanddesign/2013/feb/25/lord-rogers-parents-house-listing (accessed 26 March 2025).

54 Giles Worsley, 'Pillwood, Feock, Cornwall', *Country Life*, vol. 197, no. 32, 7 August 2003, pp. 42–5 (p. 44).

55 Duncan Macintosh, 1973, *The Modern Courtyard House*. London: Lund Humphries.

56 'The lightness of being', *Architects' Journal*, 29 September 1994, vol. 200, no. 12, pp. 27–35 (p. 33).

57 Addis and Walker 1997, p. 119.

58 'School in the sky', *Architecture Today*, no. 128, May 2002, pp. 48–56.

59 Department for Education and Skills, 2004, *Schools for the Future: Exemplar Designs – Concepts and Ideas*.

60 Ivan Harbour, 'The National Assembly for Wales – the Senedd, 1998–2004', in Sophia Psarra et al (eds), 2023, *Parliament Buildings: the Architecture of Politics in Europe*. London: UCL Press, pp. 483–91.

61 Ibid, p. 484.

62 John Winter, 'Sports Centre, Hursley, Hants', *Architectural Review*, vol. 171, no. 1021, pp. 40–1.

63 Oliver Wainwright, 'In at the deep end', *Building Design*, 12 August 2011, no. 1977, pp. 12–15 (p. 13).

64 'Nor is it yet at all improbable that suspension roofs may yet be adapted for covering large conservatories, the fulcra for the suspended chains being placed at the four corners [...] A better method still would be to extend the tubular iron columns which support the ends of the metallic gutters to a sufficient height above the valleys, and employ them as the fulcra for suspending the roof from *without*, instead of by perpendicular columns and tension rods from *within*, as has hitherto been done in roofs of extraordinary size.' Charles McIntosh, 1853, *The Book of the Garden: Vol. 1: Structural*. Edinburgh: William Blackwood and Sons, p. 114; Mark Chalmers, 'Green design', *Urban Realm*, vol. 8, no. 34, Summer 2018, pp. 82–9.

65 Alistair Fair, 2018, *Modern Playhouses: An Architectural History of Britain's New Theatres, 1945–85*. Oxford: Oxford University Press, p. 238.

66 Witold Rybczynski, 2011, *The Biography of a Building: How Robert Sainsbury and Norman Foster Built a Great Museum*. London: Thames & Hudson, p. 120.

67 *Perspecta: the Yale Architectural Journal*, vol. 7, August 1961, p. 51. Rudolph was then head of the Yale School of Architecture, where his students included Norman Foster.

68 Peter Cook, 'Sainsbury Centre for the Visual Arts', *Architectural Review*, vol. 164 no. 982, pp. 355–6.

69 Colin Amery, 1995, *Architecture, Industry and Innovation: The Early Work of Nicholas Grimshaw & Partners, 1965–88*. London: Phaidon, p. 130.

70 Tim Caulton, *Hands-On Exhibitions: Managing Interactive Museums and Science Centres*, Abingdon: Routledge, 1998, p. 8.

71 'Major office block planned for BFI IMAX site at Waterloo', 1 September 2021, https://www.se1.news/lambeth-imax-office-development/ (accessed 2 December 2024).

72 DR2007:0045, Cedric Price fonds, Canadian Centre for Architecture, Montréal; *Building Design*, no. 675, 3 February 1984, p. 2.

73 Lambeth Council, planning application 23/02218/DET, approved 8 May 2024.

74 In 1967 Venturi and Rauch proposed a giant electronic screen or 'bill-ding board' at their competition entry for the National College Football Hall of Fame, New Brunswick, New Jersey (Robert Venturi , 'A Bill-Ding-Board Involving Movies Relics and Space', *Architectural Forum*, April 1968, pp. 14–17).

75 Hugh Pearman, 'Effortless staircase keeps standards rising', RIBA *Journal*, vol. 121, no. 4, April 2013, pp. 7–9.

76 Alan Berman (ed.), 2010, *Jim Stirling and the Red Trilogy: Three Radical Buildings*. London: Frances Lincoln, p. 77.

77 Robin Nicholson, 'Where does Stirling's Olivetti sit?' A talk given to DOCOMOMO UK on 21 May 2019.

78 'Lucky Jim', *Building Design*, no. 152, 25 May 1973, pp. 12–13 (p. 13).

79 Alan Blanc et al, 1993, *Architecture and Construction in Steel*. London: Taylor & Francis, pp. 257–58.

80 Ron Herron, 'A quiet technology', https://www.pidgeondigital.com/talks/a-quiet-technology/chapters/ (accessed 27 March 2025).

81 Ron Herron, 'Imagination building, Store Street, London', AA *Files*, no. 19, Spring 1990, pp. 11–18 (p. 18).

82 Sutherland Lyall, 1992, *Imagination Headquarters: Herron Associates*, London: Phaidon, p. 4.

83 Rowan Moore, 1993, *Structure, Space and Skin: the Work of Nicholas Grimshaw & Partners*. London: Phaidon, p. 194.

84 Kenneth Powell, 2001, *Richard Rogers Complete Works: Volume 2*. London: Phaidon, p. 173.

85 Meikle, James; Pickering, Elizabeth, 'London's £550m erotic gherkin', the *Guardian*, 9 September 1996, p. 1.

86 Oliver Wainwright, 'Neo Bankside: how Richard Rogers's new "non-dom accom" cut out the poor', the *Guardian*, 21 July 2015, https://www.theguardian.com/artanddesign/architecture-design-blog/2015/jul/21/neo-bankside-how-richard-rogers-new-non-dom-accom-cut-out-the-poor (accessed 7 October 2024).

87 Estudio Cano Lasso, '125–127 Mare Street, Design and Access Statement', May 2016.

88 John Outram, 'Supermarket Forces', *Architectural Review*, vol. 173, no. 1035, May 1983, pp. 69–73 (p. 69).

89 Richard Rogers, cited in Kathryn Morrison, 2003, *English Shops and Shopping*. New Haven: Yale University Press, p. 295.

90 *Building Design*, 11 April 1975, no. 245, p. 7.

91 *Architects' Journal*, 28 September 1977, vol. 166, no. 39, p. 50.

92 W G Howell, 'Village Housing', *Architects' Journal*, vol. 126, no. 3264, 19 September 1957, pp. 428–33.

93 Amery, op. cit., p. 76.

94 Barbara Goldstein, 'Trim Tech', *Progressive Architecture*, July 1978, pp. 46–49 (p. 47).

95 *Design*, no. 380, August 1980, pp. 50–55 (pp. 50–1).

96 Annette LeCuyer, 'Refining the Shed', *Building Design*, no. 470, 2 November 1979, pp. 16–18 (p. 17).

97 It was named SSSALU: 'short span structures in aluminium' (*Architectural Review*, vol. 167, no. 995, January 1980, p. 61).

98 'Britain's Sunrise Strip', *The Economist*, 30 January 1982, pp. 78–9.

99 Derek Osbourn, 'Construction Studies 2: High-Tech Cladding', *Architects' Journal*, vol. 182, no. 36, pp. 71–82 (p. 73).

100 Amery op. cit., p. 162.

101 Amanda Birch, 'Forestry commission', *Building Design*, no. 1555, pp. 12–16 (p. 13).

102 *Architectural Review*, vol. 167, no. 1000, June 1980, pp. 376–8 (p. 376).

103 Stephen McGuckin, 'Waterloo International', *Concrete Quarterly*, no. 176, Spring 1993, pp. 18–20 (p. 20).

104 Bob Mitchell, 2003, *Jubilee Line Extension: From Concept to Completion*. London: Thomas Telford Publishing, p. 155.

glossary

AEROFOIL: a streamlined shape, like the cross-section of an airplane wing or a propeller blade, designed in architectural applications to reduce wind resistance.

BOLT-FIXED: a structural glazing system in which glass panels are held in place by bolted fixings.

BREEAM: Building Research Establishment Environmental Assessment Method, a method of assessing and certifying the sustainability of buildings.

CABLE NET: a tensile structure that relies on a network of tensioned cables spanning between fixed supports to support a membrane.

CASTELLATED: a structural steel beam with regular and repeating openings (typically hexagonal or circular) throughout its web.

CATENARY: a curve that describes the shape of a freely hanging chain or cable when supported only at its ends.

CHORD: a principal member of a truss extending from end to end.

COLD BRIDGE: a weak point in a building's insulation where heat is more readily conducted.

CROSS-LAMINATED TIMBER: a type of engineered wood created by glueing together layers of solid-sawn timber at right angles to one another.

DIAGRID (OR DIAGONAL GRID): a structural framework of diagonally intersecting members.

DUCTILE IRON: a type of cast iron where the graphite is in the form of spheres (nodules) rather than flakes, making it more ductile, stronger and tougher than traditional grey cast iron.

ETFE (ETHYLENE TETRAFLUOROETHYLENE): a fluorine-based plastic, available as a lightweight, heat-resistant film that can form a tensile membrane.

FRITTED GLASS: glass with the addition of ground-glass particles (frit) that are oven-dried and fired onto the glass, usually in a pattern.

GRC (GLASS-REINFORCED CONCRETE): a building material composed of glass fibres embedded in a concrete matrix, usually in the form of precast cladding panels.

GRIDSHELL: a lightweight grid or lattice structure that derives its strength from the interplay of its double curvature and lattice arrangement.

GRP (GLASS-REINFORCED PLASTIC): a composite material made by combining glass fibres with a polymer or resin, used to create cladding panels. Also known as fibreglass.

JACK ARCH: a shallow segmental vault spanning between beams; used in fireproof floors, etc.

K-BRACE: a pair of braces connecting to a member at mid-span, designed to take tension and compression forces.

MERO: a proprietary type of space frame (qv) developed by Max Mengeringhausen in 1943.

OPEN-WEB TRUSS: a lightweight truss comprising two parallel chords connected by triangulated web members.

PARABOLA: a U-shaped curve formed such that any point along it is equidistant from both a fixed point and a fixed straight line.

PASSIVE CHILLED BEAM: an air-conditioning device, usually suspended from the ceiling, that relies on natural convection to circulate air and provide cooling.

PLANAR GLAZING: a proprietary system of bolt-fixed structural glazing developed by Pilkington.

PORTAL FRAME: a structural frame formed of two vertical supports rigidly connected to a horizontal or cranked beam or truss, allowing the assembly to act as a single structural unit.

PRATT TRUSS: a truss with vertical web members in compression and diagonal web members in tension.

SCAGLIOLA: a type of fine plaster used in columns and other architectural elements to imitate marble.

SHELL AND CORE: the basic carcass of a building, comprising the structure and exterior envelope (or shell) and the service infrastructure (or core). In a shell and core development, these elements are completed, leaving the interior fit-out to be specified by others, including the occupant.

SLIP-CAST CONCRETE: a construction method in which concrete is poured into formwork that is continuously or incrementally raised.

SPACE FRAME (OR SPACE TRUSS): an open and lightweight structure comprising interconnected members that form a rigid three-dimensional framework.

SUPERPLASTIC (ADJ): a material that is capable of permanent deformation under load, allowing it to be easily moulded at high temperatures without fracturing.

TAKENAKA: a proprietary type of space frame consisting of two square grids rotated at 45 degrees to each other and connected with diagonal members.

TENSEGRITY (OR TENSIONAL INTEGRITY): a structural system of discontinuous compression members (such as bars or struts) held by continuous tension members (such as cables).

TRAILING EDGE: the rear edge of an aerodynamic surface such as a wing.

ULTRA HIGH-PERFORMANCE CONCRETE (UHPC): a type of concrete characterized by its high compressive strength, durability and low porosity.

further reading

Addis, B and Walker, D, 1997, *Happold: The Confidence to Build*. London: Happold Trust Publications Ltd.

Banham, R, 1960, *Theory and Design in the First Machine Age*. London: The Architectural Press.

Brenner, M, Langenberg, S, Angermann, K, and Meier, H-R, 2024, *High-Tech Heritage: (Im)permanence of Innovative Architecture*. Basel: Birkhäuser.

Davies, C, 1988, *High Tech Architecture*. London: Thames and Hudson.

Fierro, A, 2023, *Architectures of the Technopolis: Archigram and the British High Tech*. London: Lund Humphries.

Gannon, T, 2017, *Reyner Banham and the Paradoxes of High Tech*. Los Angeles: Getty Research Institute.

Harris, J B, and Pui-K Li, K, 1996, *Masted Structures in Architecture*. Oxford: Butterworth Architecture.

Jackson, N, 1996, *The Modern Steel House*. London: E & FN Spon.

Macdonald, A, 2000, *The Engineer's Contribution to Contemporary Architecture: Anthony Hunt*. London: Thomas Telford Publishing

Macdonald, A J, 2019, *High-Tech Architecture: a Style Reconsidered*. Ramsbury: The Crowood Press.

Pawley, M, 1990, *Theory and Design in the Second Machine Age*. Oxford: Basil Blackwell.

Rice, P, 1994, *An Engineer Imagines*. London: Ellipsis.

Saint, A, 2007, *Architect and Engineer: A Study in Sibling Rivalry*. New Haven: Yale University Press.

Solé Bravo, C, 2019, *Norman y Wendy Foster en Hampstead: El Sueño de la Casa Tecnológica*. Buenos Aires: Diseño.

Most of the major practices have compiled multi-volume collected works. *Richard Rogers: Complete Works* was published by Phaidon in three volumes in 1999, 2001 and 2006, edited by Ken Powell. The six-volume *Norman Foster: Works* was published by Prestel between 2003 and 2014, edited by David Jenkins. Four volumes of Nicholas Grimshaw & Partners' work have been published to date, starting with *Architecture, Industry and Innovation: the Early Work of Nicholas Grimshaw & Partners, 1965–1988*, edited by Colin Amery and published by Phaidon in 1995. Hopkins Architects have so far been the subject of three volumes, published in 1993, 2001 and 2012.

index

Photographs in **bold** type

acknowledgements

Many people have helped to make this book happen in different ways; I'm grateful to all of them. Foremost is the late Elain Harwood, whose earlier titles in this series represent a model of knowledge worn lightly. Catherine Croft and Susannah Charlton of the Twentieth Century Society provided support and encouragement, as did Charles O'Brien and James O Davies at Historic England. Alan Powers, Alistair Fair, Jon Clarke and Posy Metz made helpful comments and corrections on the draft text. Paul Catherall generously contributed cover art. My greatest thanks go to John East for his superb photography. Not only did John make his picture library available to me, he travelled far and wide to capture new images for the book.

The following are thanked for their insights on buildings or topics: Matthias Brenner and Silke Langenberg of ETH Zurich, John Chilton, Nigel Dale, Kate Davies at the Architectural Association's Hooke Park Campus, Niamh Dillon, Andy and Sophie Earl, Patty Hopkins, Elyse Howell Price, Oliver Richards and Allan Green of Orms, Billie Lee of HCLA, Robin Nicholson, , Heather Puttock of RSHP, David Rae, Ian Ritchie of Ritchie Studio, James Soames of Project Office, Jon Sykes, Amin Taha, Peter Tracey, Steve Webb of Webb Yates, Duncan Webster, Matthew Wells of Techniker Consulting Engineers, and Tom Wright and Irma Delmonte of Foster + Partners. Thanks are due to Nicola Newman, Rebecca Armstrong and colleagues at Batsford. Finally, I would like to thank Posy Metz for her support and encouragement.

picture credits

The author and publisher have made every effort to contact copyright holders and will be happy to correct, in subsequent editions, any errors or omissions that are brought to their attention.

All photographs © John East except the following: p.9 © Maison Singulier; p.11 Chicago History Museum, ICHi-151286; Raymond W. Trowbridge, photographer; p.12 © Arcaid Images / Alamy Stock Photo; p.13 Photograph by Leslie Schwartz. ©2025 Eames Office, LLC. All rights reserved; p.14 © Cedric Price fonds, Collection Centre Canadien d'Architecture / Canadian Centre for Architecture, Montréal; p.16 © Norman Foster Foundation; p.18 © Ian Dagnall / Alamy Stock Photo; p.23 Courtesy of Benthem Crouwel Architects; p.26 Photograph in the Carol M. Highsmith Archive, Library of Congress, Prints and Photographs Division; p.29 Ken Kirkwood © RSHP Archive; p.32 © Ananya Singhal; pp 36-7, p.45, p.71, p.107 © Historic England Archive / James O. Davies; p.63 © James O. Davies; p.73 © Hampshire County Council, photographer Jo Reid and John Peck; p.81 © Martine Hamilton Knight; p.85 © Geraint Franklin; p.97 © Tom McAtee / Alamy Stock Photo; p.115, p.138, p.150, p.239 © Nigel Young / Foster + Partners; p.128 © Arcaid / Richard Bryant; p.135 © Horden Cherry Lee Architects; p.149 © Historic England Archive / Chris Redgrave; p.166 © Martin Charles / RIBA Collections; p.222 Ken Kirkwood, courtesy of RSHP; p.261 © Heritage Image Partnership Ltd / Alamy Stock Photo

To the memory of Elain Harwood (1958–2023)
and Michael Franklin (1949–2024)

First published in the United Kingdom in 2026 by
Batsford
43 Great Ormond Street
London
WC1N 3HZ

An imprint of B. T. Batsford Holdings Limited

ISBN 978 1 83733 037 9

A CIP catalogue record for this book is available from the British Library.

10 9 8 7 6 5 4 3 2 1

Reproduction by Rival Colour Ltd, UK
Printed by Dream Colour, China

This book can be ordered direct from the publisher at www.batsfordbooks.com, or try your local bookshop

Distributed throughout the UK and Europe by Abrams & Chronicle Books, 1st Floor, 22–24 Ely Place, London EC1N 6TE and 57 rue Gaston Tessier, 75166 Paris, France

www.abramsandchronicle.co.uk
info@abramsandchronicle.co.uk

The title and chapter headings are set in Architype Stedelijk, based on the grid-based alphabet developed by Wim Crouwel for a 1968 poster for the Stedelijk Museum in Amsterdam. Building headings are set in Paratype's DIN 2014 and the text face is Alegreya, designed by Juan Pablo del Peral of Huerta Tipográfica.

Captions
Frontispiece (page 2): Link Centre, Swindon, 1983–5, Thamesdown Borough Council; facing contents page (page 4): Canada Water Station, Southwark, London, 1995–9, JLE Project Architects with Herron Associates.

MIX
Paper | Supporting responsible forestry
FSC® C188448
FSC www.fsc.org

C20 Twentieth Century Society

The community for modernity.

C20 Society is an independent charity and the guardian of Britain's modern design and architectural heritage.

For over 40 years, we've successfully campaigned to save countless landmarks for the nation: from iconic red phone boxes to Art Deco lidos, Brutalist bus stations to pop-art murals, even helping Bankside power station to become the cathedral of art, Tate Modern.

We believe good design enriches lives and contributes towards thriving communities, yet our shared heritage is under threat more than ever before. Each year our casework team tackle thousands of cases and help to secure listed status for remarkable buildings, while our campaigns lead the debate on the built heritage of the future, advancing environmental arguments and championing community solutions.

To find out more and get involved, visit c20society.org.uk

Together, we can protect the best of twentieth and twenty-first century architecture for future generations.